TeeJay Maths

CfE Second Level
Book 2B

Tom Strang, James Geddes, James Cairns,
Dr Naomi Norman and Catherine Murphy

Although every effort has been made to ensure that website addresses are correct at time of going to press, Hodder Education cannot be held responsible for the content of any website mentioned in this book. It is sometimes possible to find a relocated web page by typing in the address of the home page for a website in the URL window of your browser.

Hachette UK's policy is to use papers that are natural, renewable and recyclable products and made from wood grown in well-managed forests and other controlled sources. The logging and manufacturing processes are expected to conform to the environmental regulations of the country of origin.

Orders: please contact Hachette UK Distribution, Hely Hutchinson Centre, Milton Road, Didcot, Oxfordshire, OX11 7HH. Telephone: +44 (0)1235 827827. Email education@hachette.co.uk Lines are open from 9 a.m. to 5 p.m., Monday to Friday. You can also order through our website: www.hoddereducation.co.uk

ISBN: 978 1 3983 6326 7

© Thomas Strang, James Geddes, James Cairns 2023

Special thanks to Dr Naomi Norman and Catherine Murphy for their significant contribution.

First published in 2023 by

Hodder Education,
An Hachette UK Company
Carmelite House
50 Victoria Embankment
London EC4Y 0DZ

www.hoddereducation.co.uk

Impression number   10 9 8 7 6 5 4 3 2 1

Year   2027 2026 2025 2024 2023

All rights reserved. Apart from any use permitted under UK copyright law, no part of this publication may be reproduced or transmitted in any form or by any means, electronic or mechanical, including photocopying and recording, or held within any information storage and retrieval system, without permission in writing from the publisher or under licence from the Copyright Licensing Agency Limited. Further details of such licences (for reprographic reproduction) may be obtained from the Copyright Licensing Agency Limited, www.cla.co.uk

Cover illustration by Ai Higaki/D'Avila Illustration Agency

Typeset by Aptara, Inc.

Produced by DZS Grafik, Printed in Slovenia

A catalogue record for this title is available from the British Library.

Contents

Introduction		6
Chapter 0	Revision of Book 2A	8
Chapter 1	**Decimal fractions and money: Adding and subtracting tenths and hundredths**	**18**
	Adding and subtracting decimal fractions	18
	Adding and subtracting money	22
	Calculating the cost	25
	Revisit, review, revise	28
Chapter 2	**Fractions of quantities: Unit and other fractions of a quantity**	**30**
	Unit fractions of a quantity	30
	Other fractions of a quantity	33
	Revisit, review, revise	35
Chapter 3	**Whole numbers 1: Dividing with remainders and calculation practice**	**37**
	Dividing by 4 with remainders	37
	Dividing by 6 with remainders	41
	Dividing by 7 with remainders	44
	Dividing by 8 with remainders	47
	Dividing by 9 with remainders	49
	Mixed problems: more multiplication and division	52
	Revisit, review, revise	55
Chapter 4	**Sequences 1: Identifying patterns**	**56**
	Finding rules	56
	Special sequences	58
	Revisit, review, revise	61
Chapter 5	**Multiples and factors: Finding multiples and factors**	**63**
	Multiples	63
	Factors	65
	Revisit, review, revise	67
Chapter 6	**Time: Working with units of time**	**68**
	Units of time	68
	Years, months and dates	69
	Hours, minutes and seconds	73
	Timetables	75
	Longer time intervals	78
	Revisit, review, revise	80

Chapter 7	**Decimal fractions 1: Decimals to 1, 2 and 3 decimal places**	**82**
	Decimal fractions: thousandths	82
	Rounding to 1 decimal place	85
	Rounding to 2 decimal places	89
	Adding and subtracting thousandths and numbers with different decimal places	91
	Revisit, review, revise	94
Chapter 8	**Angles: Working with angles**	**95**
	Measuring and classifying angles	95
	Constructing angles	100
	Revisit, review, revise	102
Chapter 9	**Decimal fractions 2: Multiplying and dividing decimals by 10, 100 and 1000**	**103**
	Multiplying decimals by 10, 100 and 1000	103
	Dividing decimals by 10, 100 and 1000	107
	Revisit, review, revise	111
Chapter 10	**Money: Profit, loss and budgeting**	**112**
	Profit and loss	112
	Budgeting	114
	Revisit, review, revise	117
Chapter 11	**2D shapes: Triangles, squares and rectangles**	**119**
	Types of triangle	119
	Squares and rectangles	122
	Revisit, review, revise	125
Chapter 12	**Equations: Introduction to algebra**	**127**
	Greater than, less than, equal to, not equal to	127
	Simple equations	129
	Missing operations	132
	Forming equations	133
	Function machines	135
	Revisit, review, revise	139
Chapter 13	**Fractions and percentages: Understanding fractions and percentages**	**142**
	Equivalent fractions	142
	Simplifying fractions	144
	What is a percentage?	147
	Percentages and fractions	149
	Revisit, review, revise	152

Chapter 14	**Perimeter and area: Calculating perimeter and area**	**154**
	Perimeter and area: squares and rectangles	154
	Revisit, review, revise	157
Chapter 15	**3D objects and volume: 2D representation of 3D shapes**	**159**
	Volume	159
	Drawing 3D objects	162
	Nets of cubes	165
	Nets of cuboids	166
	Revisit, review, revise	168
Chapter 16	**Statistics: Understanding graphs and charts**	**170**
	Interpreting graphs and charts	170
	Misleading data	176
	Revisit, review, revise	181
Chapter 17	**Whole numbers 2: Big numbers, rounding and estimating**	**183**
	Whole numbers to 1 000 000	183
	Rounding large numbers	186
	Estimating by rounding	190
	Revisit, review, revise	193
Chapter 18	**Symmetry: Identifying and completing symmetrical shapes and patterns**	**194**
	Symmetrical shapes and patterns	194
	Revisit, review, revise	198
Chapter 19	**Sequences 2: More sequences**	**199**
	Rules for sequences	199
	Square and triangular numbers	204
	Revisit, review, revise	207
Chapter 20	**Coordinates: The coordinate axes**	**209**
	Plotting coordinates	209
	Coordinates and shapes	212
	Revisit, review, revise	216
Chapter 21	**Probability: Understanding and predicting probability**	**218**
	Probability	218
	Revisit, review, revise	221
Chapter 22	**End-of-year revision**	**222**

Answers online; scan this QR code or visit
www.hoddergibson.co.uk/teejay-second-level-maths-answers-2B

Introduction
Information for pupils

This book begins with Chapter 0, which is full of questions that revise maths you already know.

Each chapter is packed with lots of practice, covering all the maths you need to learn.

Each topic begins with a short explanation to get you started.

 I will learn how to calculate the volume of a cuboid using a formula.

There are often examples to help you understand the ideas before you answer the questions in the exercise.

Example

1 day = 24 hours $\frac{1}{4}$ day = $\frac{1}{4}$ × 24 = 6 hours

Use these questions at the end of each chapter to look back at what you have learned.

Revisit, review, revise

1 Choose a word from the box to describe each angle

| acute | right angle | obtuse | straight | reflex |

Now try this!

Play-based activities help you learn while you have fun.

These yellow boxes introduce you to some new maths.

Remember, remember

Sometimes, you need to remember some maths that you have learned before.

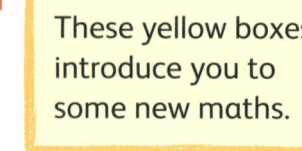

Review what you have learned in the end-of-year chapter at the end of the book.

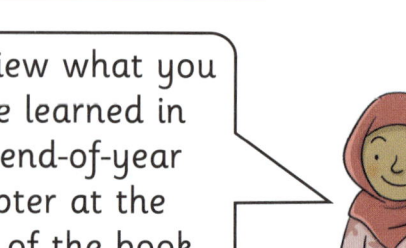

Introduction

Information for teachers, parents and carers

Welcome to the second edition of a well-loved TeeJay Maths series. The Second Level scheme has been restructured so that it comprises three books instead of two, in line with curriculum structure. Book 2A covers the course for P5, Book 2B covers P6 and Book 2C covers P7.

Many of the familiar TeeJay features have been retained, including a **Chapter 0** at the start of each book, which revisits topics learned at the previous level. Additionally, each chapter ends with a **'Revisit, review, revise'** section and each book ends with an **End-of-year revision** chapter.

Progression is built into the structure of each book, with Whole Number chapters occasionally interchanging with other topics. Questions for differentiation have been flagged throughout:

- Easier questions/activities, or building blocks, are flagged by this icon
- Hard questions, or stretch, are flagged by this icon

Activities for **play-based learning** (Now try this!) have been embedded throughout to engage pupils in their learning.

Answers to all questions can be found by scanning the QR code below or online at: www.hoddergibson.co.uk/teejay-second-level-maths-answers-2B

In addition to the three second edition textbooks, new **interactive resources, editable course plans, teaching guides** and **worksheets** will be available through our Boost platform. The worksheets include **mental maths, practice** (to test pupils on each unit) and **assessment** (to be taken at half-term). The practice and assessment worksheets are available in digital and PDF format. The mental maths are PDF only.

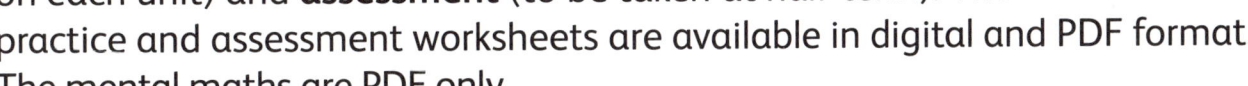

www.hoddergibson.co.uk/teejay-second-level-maths-boost

0 Revision of Book 2A

1) Match the pictures to these fractions: $\frac{1}{2}$ $\frac{1}{3}$ $\frac{1}{4}$

   a) 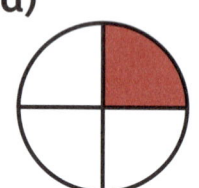 b) 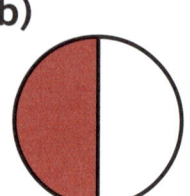 c)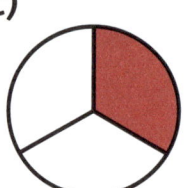

2) | 7481 | 3470 | 9716 | 1087 |

   Which number from the above list has:
   a) 7 hundreds
   b) 7 ones
   c) 7 thousands
   d) 7 tens?

3) | 20.6 | 6.02 | 62 | 60.2 |

   Which number has:
   a) 2 tenths
   b) 2 tens
   c) 2 hundredths
   d) 2 ones?

4) In your jotter, write **five thousand and fourteen** in digits.

5) In your jotter, write these numbers in order of size.

   | 1043 | 431 | 4310 | 1304 | 1034 |

   Start with the **biggest**.

Revision of Book 2A

6) Work out:

a) 263 + 409

b) 4261 + 3958

c) 658 − 132

d) 796 − 508

e) 5284 − 2916

f) 8641 − 953

7) Round:
   a) 6.9 to the nearest whole number
   b) 52.3 to the nearest whole number
   c) 101.5 to the nearest whole number.

8) What is:
   a) 6 × 5
   b) 7 × 2
   c) 3 × 9
   d) 8 × 10
   e) 2 × 6
   f) 6 × 7
   g) 9 × 5
   h) 8 × 8
   i) 8 × 3
   j) 7 × 4
   k) 4 × 9
   l) 6 × 8

9) What are the **missing** numbers?
   a) 4 × ___ = 36
   b) 8 × ___ = 48
   c) 9 × ___ = 72

10) Work out:
    a) 22 × 3
    b) 46 × 2
    c) 97 × 4

11) Work out:
    a) 8 × 1000
    b) 340 × 10
    c) 19 × 100

12) Work out:
    a) 2000 ÷ 1000
    b) 9500 ÷ 10
    c) 4100 ÷ 100

Chapter 0 Revision of Book 2A

13) In your jotter, write these amounts:
   a) £4.56 in pence
   b) £0.87 in pence
   c) 596p in £
   d) 91p in £

14) Three children, Ann, Ben and Cara, share 15 sweets.

      Ann   Ben   Cara

Do they each have a third of the sweets?

In your jotter, write **yes** or **no**.

15) Is $\frac{1}{5}$ smaller than $\frac{1}{8}$? Yes or no.

Use the picture to help you.

16) In your jotter, complete the equivalent fractions:

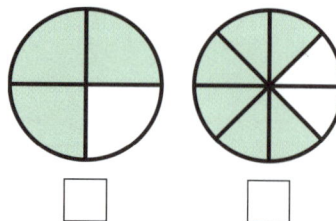

$$\frac{\square}{4} = \frac{\square}{8}$$

Revision of Book 2A

17) In your jotter, write the decimals that the arrows point to.

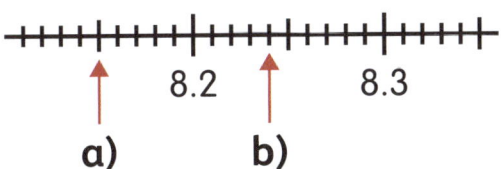

a)   b)

18) Work out:

a) 6) 66   b) 4) 484   c) 7) 84

d) 8) 968   e) 9) 927   f) 6) 822

For questions 19, 20, 21 and 22, you must decide whether to add, subtract, multiply or divide.

19) A pack of 4 pens costs 76p.
How much is 1 pen?
Give your answer as £____.____

20) 1637 people visit a shopping centre.
465 people arrive by bus.
How many do **not** arrive by bus?

21) In a cupboard there are 6 shelves of books.
There are 24 books on each shelf.
How many books in the cupboard?

22) How many lines of symmetry does each shape have?

a)    b)    c)

Chapter 0 Revision of Book 2A

23) Trace or copy this shape.

Complete the shape so that the dashed line is a line of symmetry.

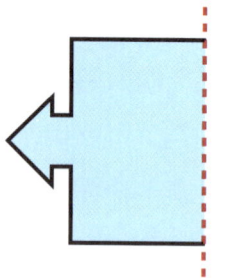

24) What times are shown?

In your jotter, write the answers in **digits** (use a.m. or p.m.) and in **words**.

a)  in the evening

b)  in the morning

25) Change the 12-hour clock times to 24-hour clock times.
    a) 7:30 a.m.
    b) 4:25 p.m.
    c) midday
    d) 12:35 a.m.

26) Change the 24-hour clock times to 12-hour clock times, using a.m. or p.m.
    a) 22:15
    b) 06:00
    c) 10:07
    d) 00:00

27) How long is it from 9:15 a.m. to 11:30 a.m.?

28) Copy or trace triangle PQR.

Colour:

a) <PQR yellow

b) <QRP red.

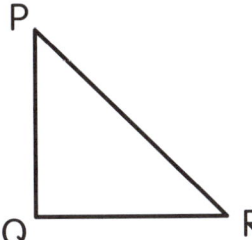

29) Choose a word from the list to describe each angle.

| acute | obtuse | reflex | straight | right angle |

a)

b)

c)

d)

e)

30) How many degrees from:

a) East to South (clockwise)

b) North-West to South (anti-clockwise)?

Chapter 0 Revision of Book 2A

31) This polygon has 6 sides.
   a) How many angles does it have?
   b) What is the name of this shape?

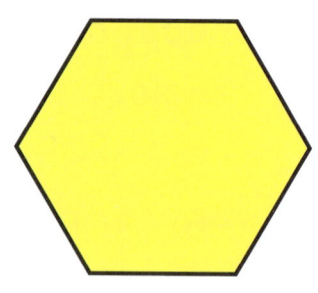

32) In your jotter, write the name of each shape. Choose from the list.

| square   rectangle   rhombus   parallelogram   trapezium   kite |

a)   b)   c)

d)   e)   f)

33) 1 cm = 10 mm, 1 m = 100 cm and 1 km = 1000 m.
   a) How many mm in 8 cm?
   b) How many cm is 60 mm?
   c) How many m is 200 cm?
   d) How many cm in 4 m?
   e) How many km is 7000 m?

34) The triangle has perimeter 68 cm. What is the **missing** length?

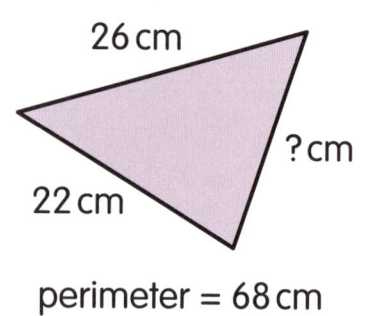

35) Calculate the area of each rectangle.

a)

b)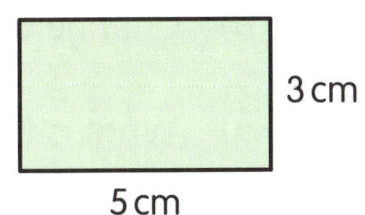

Each square is 1 cm × 1 cm

36) What is the volume of this shape in cubic cm (cm³)?

Each cube is 1 cm × 1 cm × 1 cm

37) What is the volume of liquid, in millilitres, in this bottle?

38) a) Change 4000 millilitres to litres.

b) Change 6 litres 750 millilitres to millilitres.

39) a) What is this shape called?

b) How many faces does it have?

c) What shape is each face?

d) How many vertices (corners) does it have?

e) How many edges does it have?

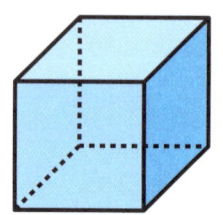

40) Name each 3D shape.

a)    b) 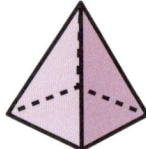   c)    d)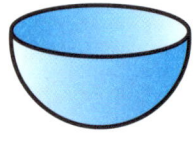

41) a) In your jotter, write 4 kg 58 g in grams.

b) Change 6923 g to kilograms and grams.

42) Draw an 8 × 8 coordinate grid as shown.

a) Plot the points **W** (1, 3), **X** (7, 3) and **Y** (7, 7).

b) Join W → X → Y.

c) Plot the point **Z** to make a rectangle.

d) What are the coordinates of Z?

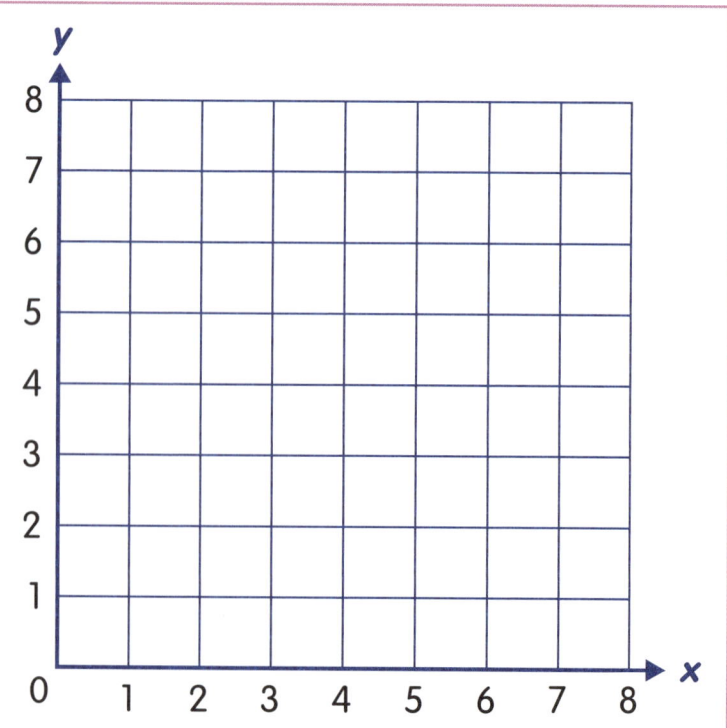

Revision of Book 2A

43) The pictograph shows the number of times Priya played football each month.

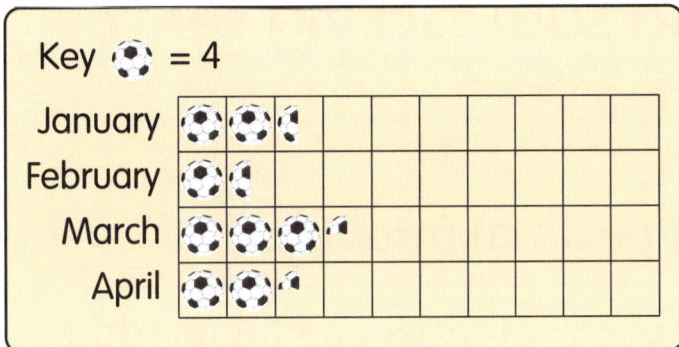

How many times did she play:
a) in January
b) in April
c) in March
d) altogether?

44) The number of bookings at a restaurant are shown in the bar chart.
a) How many bookings on Thursday?
b) On what day were there the **most** bookings?
c) How many **more** bookings on Wednesday than on Monday?
d) How many bookings **altogether**?

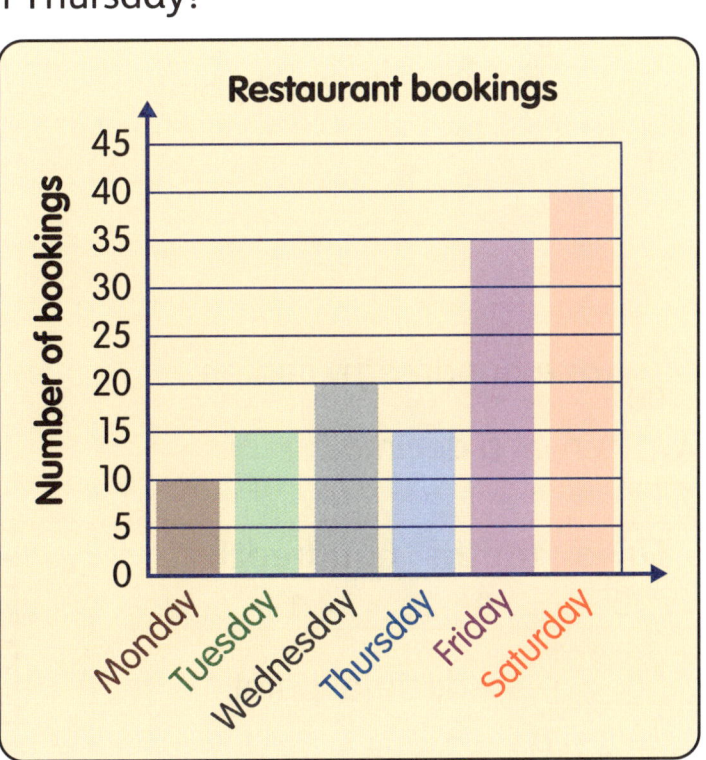

1 Decimal fractions and money

Adding and subtracting tenths and hundredths

Adding and subtracting decimal fractions

 I will learn to add and subtract tenths and hundredths.

Remember, remember

$\frac{1}{10}$ (one tenth) can be written as a decimal:

Ones	.	tenths
0	.	1

— decimal point

$\frac{1}{10} = 0.1$

10 tenths = 1 one

$\frac{10}{10} = 1$

$\frac{1}{100}$ (one hundredth) can be written as a decimal:

Ones	.	tenths	hundredths
0	.	0	1

— decimal point

$\frac{1}{100} = 0.01$

10 hundredths = 1 tenth

$\frac{10}{100} = \frac{1}{10}$

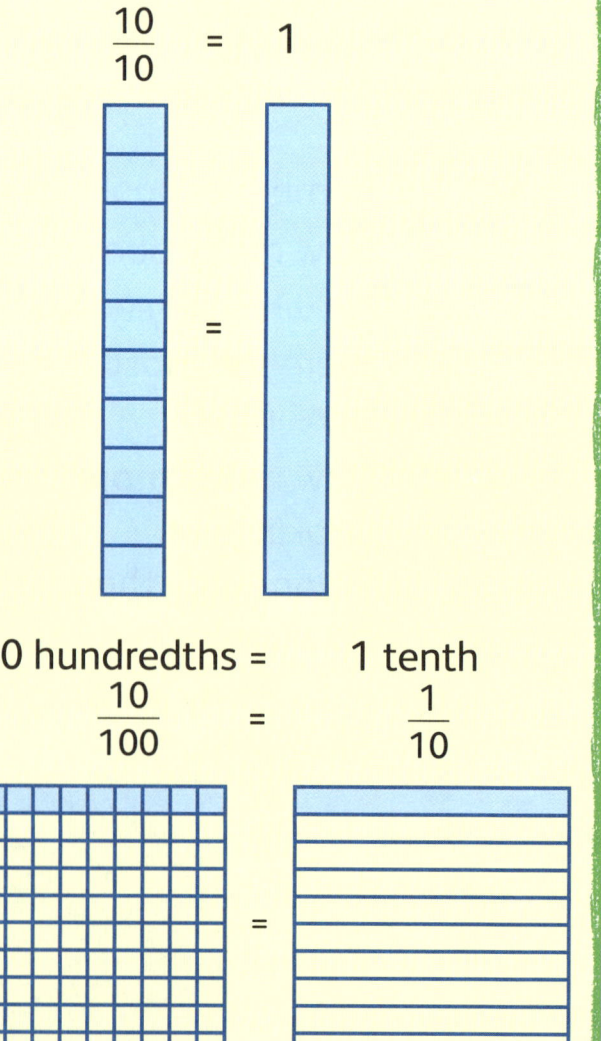

18

Adding and subtracting tenths and hundredths

You can **add and subtract decimals mentally**.

Look at which place value will change by the addition or subtraction.

For example:

6.2 + 0.3 =

You are adding tenths. Only the number of tenths changes.

6 ones + 2 tenths + 3 tenths = 6 ones + 5 tenths
= 6.5

54.87 − 0.04 =

You are subtracting hundredths. Only the number of hundredths changes.

5 tens + 4 ones + 8 tenths + 7 hundredths − 4 hundredths
= 5 tens + 4 ones + 8 tenths + 3 hundredths
= 54.83

You can **add and subtract decimals using place value columns**.

For example:

	Hundreds	Tens	Ones	.	tenths	hundredths
	5	4	2	.	3	7
+	4	2	7	.	2	1
	9	6	9	.	5	8

> When adding, line up the digits with the same place value.

Sometimes you may have to carry when adding or exchange when subtracting.

For example:

	Tens	Ones	.	tenths
	9	³4̸	.	¹2
−	3	1	.	8
	6	2	.	4

← You cannot subtract 8 tenths from 2 tenths

Exchange 1 one for 10 tenths and carry them into the tenths column
Now there are 3 ones left and 12 tenths − 8 tenths

Chapter 1 Decimal fractions and money

Exercise 1

 1) Copy and complete to answer these questions mentally (without writing any working):

a) 4.6 + 0.2 = 4 ones + 6 tenths + 2 tenths = 4.___

b) 25.7 − 0.1 = 2 tens + 5 ones + 7 tenths − 1 tenth = 25.___

c) 18.45 − 0.03
= 1 ten + 8 ones + 4 tenths + 5 hundredths − 3 hundredths
= 18.4___

d) 4.13 + 0.8 = 4 ones + 1 tenth + 3 hundredths + 8 tenths
= 4.___3

2) Mentally work out the number that is:

a) 0.01 more than 1.34 b) 0.3 more than 15.6

c) 0.6 more than 326.2 d) 0.5 less than 36.7

 e) 0.11 less than 78.43 f) 108.3 more than 0.7

3) Mentally work out:

a) 8.3 + 0.4 b) 19.6 + 0.2

c) 71.8 − 0.5 d) 184.9 − 0.3

e) 2078.5 − 0.1 f) 89.04 − 0.01

g) 754.43 − 0.02 h) 846.92 − 0.31

4) Work out:

a) 31.6 + 28.2 b) 405.3 + 12.8 c) 8.95 + 1.03

d) 14.62 + 63.18 e) 1045.8 + 2931.5 f) 382.69 + 243.55

Adding and subtracting tenths and hundredths

5) Line up the digits with the same place value to work out these additions:
   a) 1.9 + 8.4
   b) 3.34 + 5.45
   c) 1.49 + 8.37
   d) 10.41 + 3.78
   e) 3.27 + 122.44
   f) 4051.2 + 50.5
   g) 1093.14 + 2.95
   h) 103.99 + 4.33
   i) 982.73 + 9.68

6) Work out:
   a) 53.7 − 21.4
   b) 516.4 − 102.5
   c) 6.08 − 3.27
   d) 25.74 − 12.37
   e) 2149.2 − 315.1
   f) 567.19 − 218.68

7) Line up the digits with the same place value to work out these subtractions:
   a) 8.4 − 5.7
   b) 18.3 − 1.6
   c) 7.65 − 5.31
   d) 29.72 − 18.37
   e) 137.44 − 3.27
   f) 6536.2 − 10.5
   g) 9512.14 − 8.05
   h) 184.25 − 1.69
   i) 831.72 − 5.84

8) In an ice skating competition, Marion is awarded 1.9 points more than Kamal.

   Kamal receives 9.4 points.

   How many points does Marion receive?

9) One Monday in May, 12.15 mm of rain fell in the Scottish Highlands.

   On the same day, 0.29 mm of rain fell in Pisa, Italy.

   How much **more** rain fell in the Scottish Highlands?

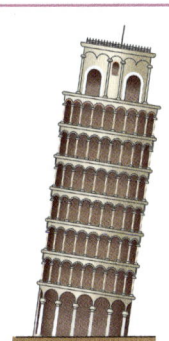

Chapter 1 Decimal fractions and money

 10) Box A has a mass of 41.72 kilograms.
Box B is 38.98 kilograms lighter.
What is the **total mass** of Box A and Box B?

Adding and subtracting money

 I will learn to add and subtract money.

Remember, remember

100p = £1

This means 1p = £$\frac{1}{100}$ = £0.01

Pounds and pence are often written with a **decimal point**.

When working with money, you always have two numbers to the right of the decimal point.

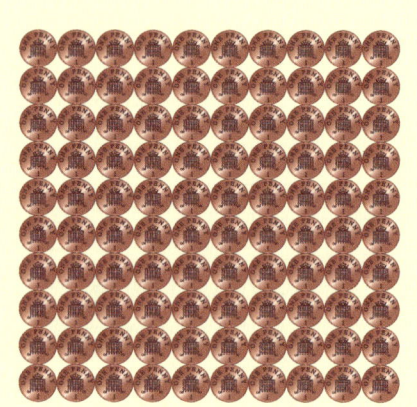

You can **add and subtract money mentally**.
Look at which place value will change by the addition or subtraction.
For example:

£5.23 + £0.70 =

You are adding tenths. Only the number of tenths changes.

5 ones + 2 tenths + 3 hundredths + 7 tenths
= 5 ones + 9 tenths + 3 hundredths
= £5.93

Adding and subtracting tenths and hundredths

£26.39 – £0.08 =

You are subtracting hundredths. Only the number of hundredths changes.

2 tens + 6 ones + 3 tenths + 9 hundredths – 8 hundredths
= 2 tens + 6 ones + 3 tenths + 1 hundredth
= £26.31

You can **add and subtract money using place value columns.**

Example 1

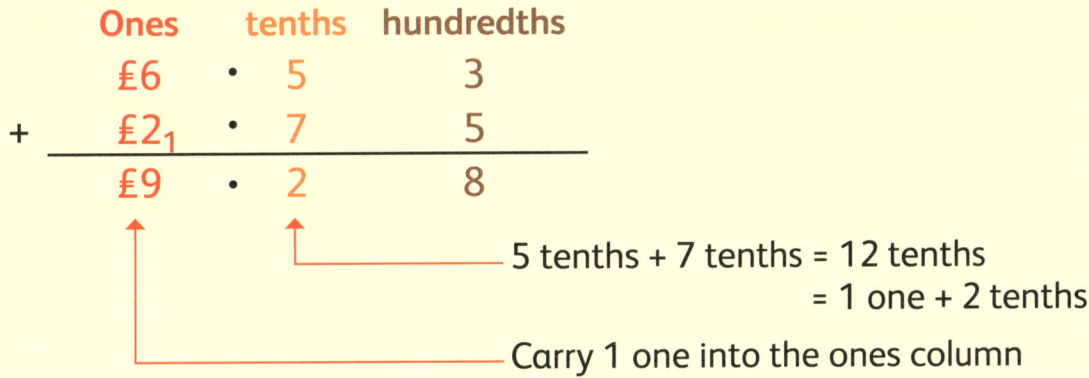

5 tenths + 7 tenths = 12 tenths
= 1 one + 2 tenths

Carry 1 one into the ones column

Example 2

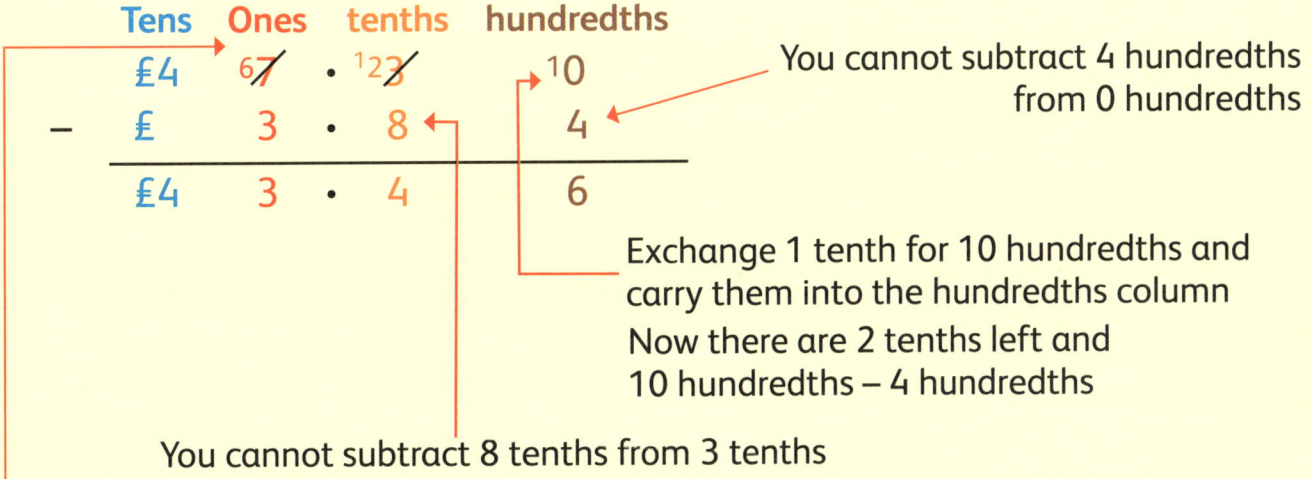

You cannot subtract 4 hundredths from 0 hundredths

Exchange 1 tenth for 10 hundredths and carry them into the hundredths column
Now there are 2 tenths left and
10 hundredths – 4 hundredths

You cannot subtract 8 tenths from 3 tenths

Exchange 1 one for 10 tenths and carry them into the tenths column
Now there are 6 ones left and 12 tenths – 8 tenths

Chapter 1 Decimal fractions and money

Exercise 2

 1) Copy and complete to answer these questions mentally (without writing any working):
a) £7.50 + £0.30 = £7.___0
b) £11.64 − £0.50 = £11.___4
c) £16.83 + £0.02 = £16.8___
d) £751.99 − £0.09 = £751.9___

2) Mentally work out the amount that is:
a) £0.03 more than £5.62
b) £0.40 more than £28.50
c) £0.06 less than £34.27
d) £0.60 less than £924.78
e) £0.61 less than £12.79
f) £0.60 more than £78.40

3) Work out:

a) £2.51
  + £1.28

b) £5.83
  − £1.72

c) £46.84
  + £32.18

d) £128.78
  + £713.92

e) £14.82
  − £ 9.28

f) £405.75
  − £120.86

4) Line up the digits with the same place value to work out these amounts:
a) £6.15 + £2.74
b) £7.58 − £5.22
c) £12.93 + £56.21
d) £68.43 − £24.06
e) £315.86 − £102.37
f) £9325.97 − £1006.88
g) £9.56 + £109.78
h) £12.00 − £10.89

24

Adding and subtracting tenths and hundredths

Calculating the cost

> I will learn to calculate costs involving adding and subtracting money.

Remember, remember

When you do not have the exact money for an item, you may give too much.

Then you get money back.

This is called **change**.

You can use a number line to work out change.

For example:

Ali buys a baseball cap for £6.79. He pays with a £10 note.

Count on from £6.79 to £10 in hundreds of pennies (£1s), tens (10ps), then ones (1ps).

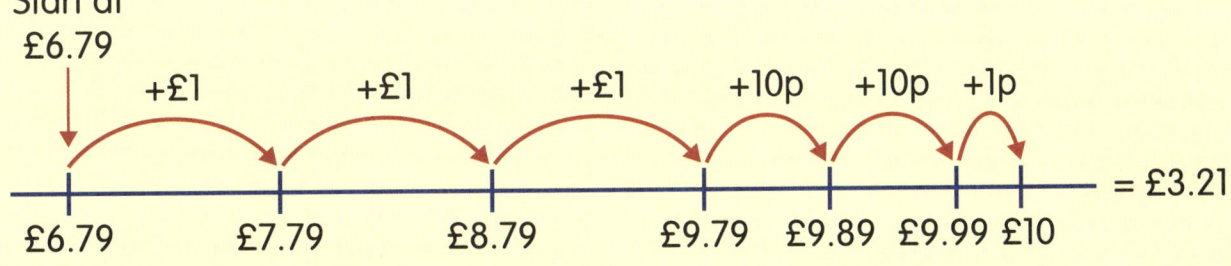

Ali gets £3.21 change.

You may choose to answer a money problem involving addition or subtraction:
- mentally
- using a number line
- using column addition or subtraction.

Chapter 1 Decimal fractions and money

Sometimes a problem may have one cost in pounds and one in pence. For example:

Lee buys a box of biscuits for £2.65 and a packet of biscuits for 76p.
You can work out the **total** he spends using a **number line**.

Start at £2.65

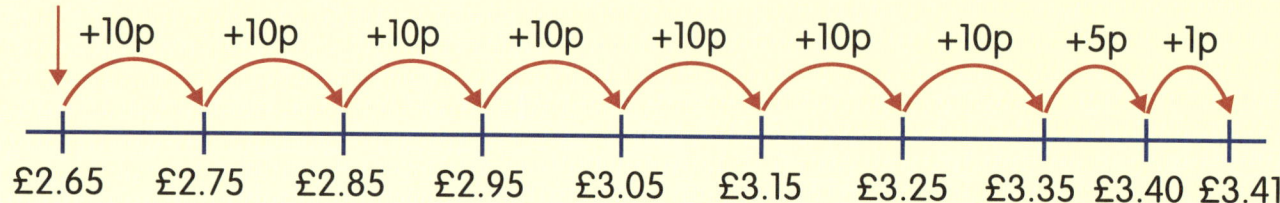

Or you can work out the total using **column addition**.
To do this, you must change **both amounts to pounds** with a decimal point.

$$\begin{array}{r} £2\,.6\,5 \\ +\ £0_1.7_16 \\ \hline £3\,.4\,1 \end{array}$$

In total, Lee spends £3.41

Exercise 3

Answer these money problems mentally, using a number line, or using column addition or subtraction.

1) Rose pays £6.99 for a book on her e-reader.
   She also pays £10 to download some music.
   How much does Rose pay **altogether**?

Adding and subtracting tenths and hundredths

2) Mo has £5.90 in his wallet.
   He buys a parking ticket for 40p.
   How much does Mo have **left**?

3) Jayu buys a necklace for £45.50 and earrings for £12.65
   How much does she spend **altogether**?

4) A lawnmower is £89.65 in the garden centre.
   The same lawnmower is £29.89 **less** online.
   How much is the lawnmower online?

5) A new camera costs £382.35
   A second-hand one costs £223.70
   What is the difference in price?

6) A family holiday for one week costs £2098.36
   A second week is another £1982.95
   How much does the two week holiday cost in **total**?

7) Paul and Sue are saving to buy a car.
   Paul has £3065.90 in his savings account.
   Sue has £4018.85 in her savings account.
   The car they want to buy is £7500
   How much **more** money do they need to save?

Chapter 1 Decimal fractions and money

☀ 8) This week's offers at a local shop are:

Jaya has £5

Does she have enough to buy two toilet rolls, a comb and a toothbrush?

Show your working.

Now try this!

A caterpillar hatches on 1st June.

The next day it grows 0.2 cm.

Then each day after that it grows 0.2 cm.

On 5th June, it is 1.8 cm.

How long is it when it hatches?

How long is it on 9th June?

Challenge

When it reaches 5.4 cm it becomes a butterfly.

How many days old is the caterpillar when it changes into a butterfly?

Revisit, review, revise

1) Mentally work out:
   - a) 3.8 + 6.1
   - b) 9.5 − 0.4
   - c) 0.92 − 0.3
   - d) 6.8 + 1.2

Adding and subtracting tenths and hundredths

2) Mentally work out:
   a) £8.32 + £0.02
   b) £3.65 + £0.40
   c) £641.04 – £0.01
   d) £76.89 – £0.60

3) Mentally work out the number that is:
   a) 0.7 more than 1.2
   b) 0.08 less than 14.99

4) Work out:
   a) 46.8 + 34.1
   b) 108.76 + 521.53
   c) 937.6 + 203.8

5) Work out:
   a) £3.95 + £2.82
   b) £563.24 + £35.19
   c) £3647.50 + £1422.86

6) Line up the digits with the same place value to work out:
   a) 9.6 – 3.4
   b) 10.3 + 4.6
   c) 947.6 – 152.7
   d) 805.2 + 296.9
   e) £18.45 + £5.17
   f) £2057.34 – £1649.82

7) The cost of 1 adult and 1 child for a ride at the fair is £11.50
   An adult ticket is £8.30
   How much is a child's ticket?

8) Alice is 1.87 m tall.
   Kofi is 0.93 m tall.
   How much **taller** is Alice than Kofi?

9) Josh buys a tennis racket for £46.70 and tennis balls for £6.35
   a) How much does he pay in **total**?
   b) He pays with three £20 notes.
      How much **change** does he get?

10) Two numbers added together total 13.45
    One of the numbers is 7.62
    What is the other number?

2 Fractions of quantities
Unit and other fractions of a quantity

Unit fractions of a quantity

💡 I will learn to find a unit fraction of an amount.

Remember, remember

Finding $\frac{1}{2}$ (**half**) is the same as **sharing equally between 2** or **dividing by 2**.

Finding $\frac{1}{3}$ (**third**) is the same as **sharing equally between 3** or **dividing by 3**.

All fractions with a numerator (top number) of **1** are called **unit fractions**. For example, $\frac{1}{2}, \frac{1}{3}, \frac{1}{4}$ and $\frac{1}{5}$ are unit fractions.

Finding $\frac{1}{4}$ (**quarter**) is the same as **sharing equally between 4** or **dividing by 4**.
For example:

$\frac{1}{4}$ of 12 = 12 ÷ 4
   = 3

or

Unit and other fractions of a quantity

Finding $\frac{1}{5}$ (**fifth**) is the same as **sharing equally between 5** or **dividing by 5**.

For example:
$\frac{1}{5}$ of 10 = 10 ÷ 5
       = 2

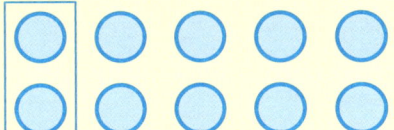 or

Remember, remember

Knowing your times tables helps when dividing.

For example:

4 × 3 = 12	so 12 ÷ 3 = 4	5 × 2 = 10	so 10 ÷ 2 = 5
3 × 4 = 12	so 12 ÷ 4 = 3	2 × 5 = 10	so 10 ÷ 5 = 2

Exercise 1

 1) Use the array to help you find the unit fractions.

a) $\frac{1}{4}$ of 20

b) $\frac{1}{5}$ of 20

 2) Use the bar to help you find $\frac{1}{6}$ of 12.

Chapter 2 Fractions of quantities

3) Copy and complete:

a) $\frac{1}{4}$ of 8 = 8 ÷ 4 = ____

b) $\frac{1}{5}$ of 50 = ____ ÷ 5 = ____

c) $\frac{1}{6}$ of 18 = ____ ÷ ____ = ____

d) $\frac{1}{9}$ of 45 = ____ ÷ ____ = ____

4) Find:

a) $\frac{1}{5}$ of 15

b) $\frac{1}{4}$ of 24

c) $\frac{1}{8}$ of 40

d) $\frac{1}{6}$ of 36

e) $\frac{1}{7}$ of 28

f) $\frac{1}{8}$ of 72

5) There are 48 jelly beans in a jar.
$\frac{1}{8}$ of them are red.
How many jelly beans are red?

6) Jerry drives 35 km.
After driving $\frac{1}{5}$ of this distance, he stops for petrol.
How far into his trip does he stop for petrol?

7) There are 60 trees in an orchard.
$\frac{1}{10}$ are apple trees.
How many are **not** apple trees?

8) Ali makes 42 sandwiches for a party.
$\frac{1}{3}$ of them are filled with cheese.
How many cheese sandwiches does Ali make?

Unit and other fractions of a quantity

Other fractions of a quantity

 I will learn to find other fractions of an amount.

To find any fraction of an amount, **find the unit fraction and then multiply**.

Example

Find $\frac{4}{5}$ of 15.

Divide by 5 to find $\frac{1}{5}$ (one fifth): $\frac{1}{5}$ of 15 = 15 ÷ 5 = 3

 or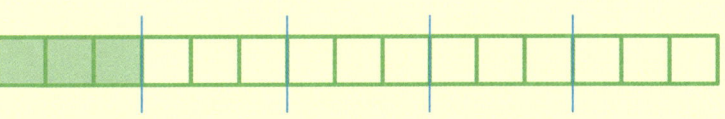

Then multiply by 4 to find $\frac{4}{5}$ (four fifths): 3 × 4 = 12

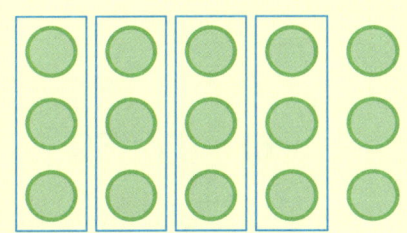 or

Exercise 2

 1) Use the array to help you find:
   a) $\frac{1}{3}$ of 18
   b) $\frac{2}{3}$ of 18
   c) $\frac{1}{6}$ of 18
   d) $\frac{5}{6}$ of 18

Chapter 2 Fractions of quantities

2) Use the bar to help you find $\frac{3}{4}$ of 12.

3) Copy and complete:

   a) $\frac{2}{3}$ of 30

   $\frac{1}{3}$ of 30 = 30 ÷ ___ = ___

   $\frac{2}{3}$ of 30 = ___ × 2 = ___

   b) $\frac{3}{4}$ of 12

   $\frac{1}{4}$ of 12 = ___ ÷ ___ = ___

   $\frac{3}{4}$ of 12 = ___ × 3 = ___

   c) $\frac{5}{6}$ of 24

   $\frac{1}{6}$ of 24 = ___ ÷ ___ = ___

   $\frac{5}{6}$ of 24 = ___ × 5 = ___

   d) $\frac{6}{7}$ of 35

   $\frac{1}{7}$ of 35 = ___ ÷ ___ = ___

   $\frac{6}{7}$ of 35 = ___ × ___ = ___

4) Find:

   a) $\frac{3}{4}$ of 20

   b) $\frac{2}{3}$ of 24

   c) $\frac{5}{6}$ of 60

   d) $\frac{7}{10}$ of 70

   e) $\frac{4}{9}$ of 27

   f) $\frac{3}{8}$ of 48

5) Emilie has 36 vegetable seeds.
   $\frac{5}{9}$ of them are beetroot seeds.
   How many beetroot seeds does she have?

6) A baker makes 54 rolls.
   $\frac{5}{6}$ of them have seeds on them.
   How many rolls have seeds on them?

Unit and other fractions of a quantity

 7) Miguel has £24.

He gives $\frac{5}{8}$ to his brother.

The rest he keeps for himself.

How much does he keep for himself?

Now try this!

Ash says:

I am thinking of a number.
It is bigger than 5 and smaller than 25.
I can find $\frac{1}{2}$ of my number with **no remainder**.
I can find $\frac{1}{4}$ of my number with **no remainder**.
When I find $\frac{1}{6}$ of my number there is a **remainder of 2**.

Work with a partner to find Ash's number.

Revisit, review, revise

1) Copy and complete:

   a) $\frac{1}{10}$ of 80 = 80 ÷ 10 = ____

   b) $\frac{1}{5}$ of 45 = ____ ÷ 5 = ____

   c) $\frac{1}{4}$ of 32 = ____ ÷ ____ = ____

   d) $\frac{1}{6}$ of 42 = ____ ÷ ____ = ____

   e) $\frac{1}{9}$ of 54 = ____ ÷ ____ = ____

   f) $\frac{1}{7}$ of 28 = ____ ÷ ____ = ____

Chapter 2 Fractions of quantities

2) Find:

   a) $\frac{1}{5}$ of 25   b) $\frac{1}{4}$ of 40   c) $\frac{1}{8}$ of 64   d) $\frac{1}{6}$ of 60

   e) $\frac{1}{10}$ of 30   f) $\frac{1}{9}$ of 63   g) $\frac{1}{7}$ of 49   h) $\frac{1}{6}$ of 54

3) Copy and complete:

   $\frac{7}{8}$ of 32:

   $\frac{1}{8}$ of 32 = ___ ÷ ___ = ___

   $\frac{7}{8}$ of 32 = ___ × ___ = ___

4) Find:

   a) $\frac{2}{3}$ of 9   b) $\frac{4}{5}$ of 30   c) $\frac{2}{7}$ of 42   d) $\frac{2}{9}$ of 81

   e) $\frac{3}{10}$ of 40   f) $\frac{7}{8}$ of 80   g) $\frac{8}{9}$ of 72   h) $\frac{3}{7}$ of 63

5) There are 28 children in a class. $\frac{3}{4}$ of them go to music club.

   How many go to music club?

6) A game has 40 cards.

   a) $\frac{2}{5}$ are red. How many are red?

   b) $\frac{3}{10}$ are blue. How many are blue?

   c) The rest are yellow. How many are yellow?

3 Whole numbers 1
Dividing with remainders and calculation practice

Dividing by 4 with remainders

 I will learn to divide by 4 when there is a remainder.

Remember, remember

Dividing by 4 is the same as **sharing equally between 4** or **grouping in 4s**.

12 buttons are **shared equally** between 4 children: 12 ÷ 4 = 3

There are 3 **groups of 4** in 12 or 4 into 12 goes 3 times: 12 ÷ 4 = 3

You can write a division using ÷ or ⌐

Sometimes, when you **divide** or **share**, there are some left over.

This is called a **remainder**.

Share 14 buttons between 4 children.

Each child gets 3 buttons.

There are 2 buttons left over.

There is a **remainder of 2**.

```
   3 r 2
4 ⟌ 1 4
```

14 ÷ 4 = 3 r 2

Chapter 3 Whole numbers 1

Example

Example 1

Work out 85 ÷ 4.

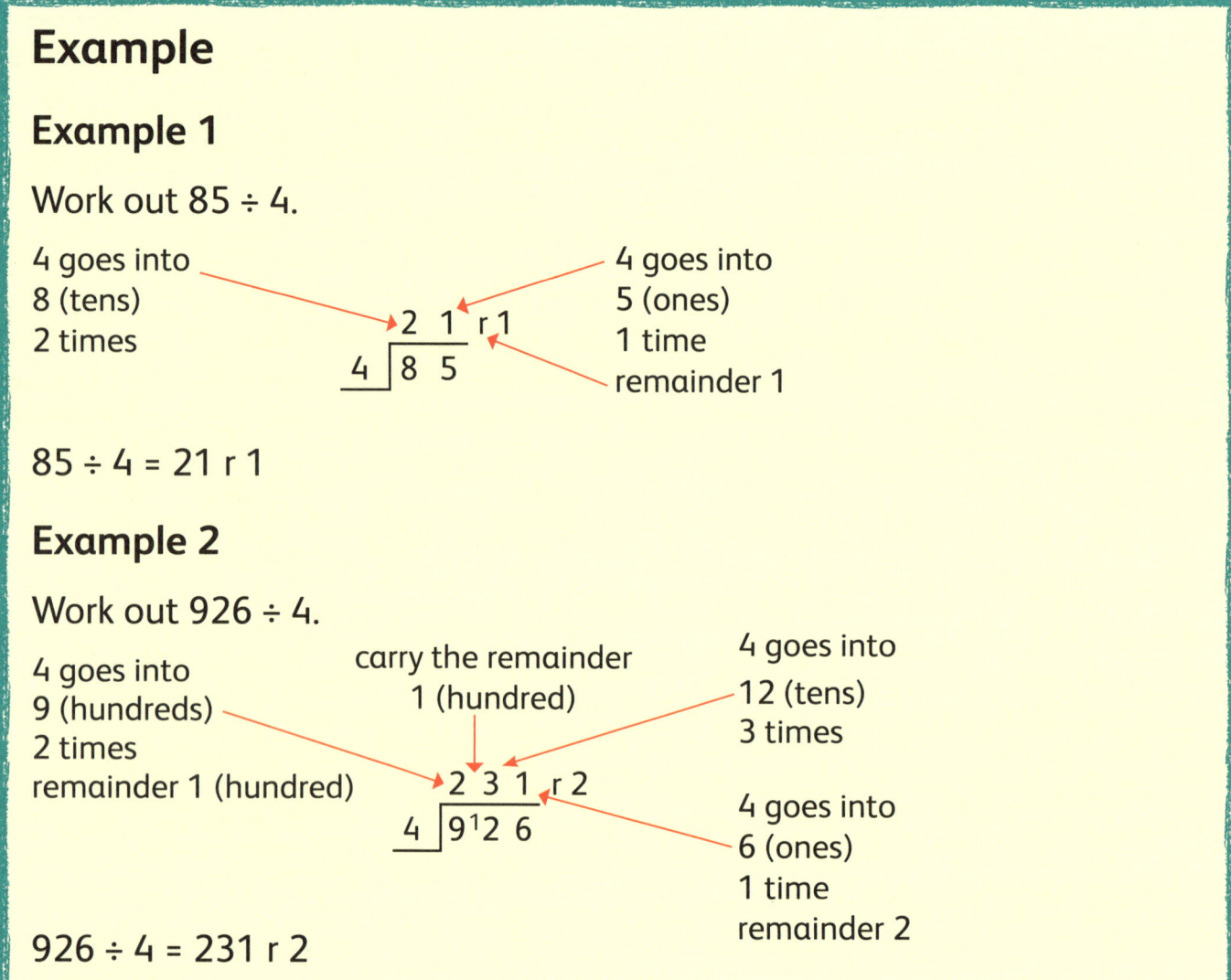

4 goes into 8 (tens) 2 times

4 goes into 5 (ones) 1 time remainder 1

85 ÷ 4 = 21 r 1

Example 2

Work out 926 ÷ 4.

4 goes into 9 (hundreds) 2 times remainder 1 (hundred)

carry the remainder 1 (hundred)

4 goes into 12 (tens) 3 times

4 goes into 6 (ones) 1 time remainder 2

926 ÷ 4 = 231 r 2

Exercise 1

1) Knowing your 4 times table helps when dividing by 4.
   In your jotter, write the 4 times table up to 4 × 9.

Dividing with remainders and calculation practice

 2) Work out these divisions.
You can use cubes or counters to help you.

a) 5 ÷ 4

b) 7 ÷ 4

c) 9 ÷ 4

d) 10 ÷ 4

3) Copy and complete:

a) 4) 8 2 = ☐ 0 r ☐

b) 4) 6 ²4 = ☐☐

c) 4) 7 ³5 = ☐☐ r ☐

d) 4) 5 ¹6 8 = ☐☐☐

e) 4) 8 7 ³3 = ☐☐☐ r ☐

f) 4) 9 ¹5 ³6 = ☐☐☐

g) 4) 4 3 ³2 = ☐ 0 ☐

h) 4) 8 1 ¹4 = ☐☐☐ r ☐

4) Work out:

a) 4) 68

b) 4) 85

c) 4) 71

d) 4) 104

e) 4) 968

f) 4) 249

g) 4) 536

h) 4) 937

Chapter 3 Whole numbers 1

5) In your jotter, write each division using ⌐. Then work out the answer.

    a) 76 ÷ 4   **b)** 51 ÷ 4   **c)** 847 ÷ 4   **d)** 458 ÷ 4

6) 49 books are packed equally into 4 boxes.
    a) How many books in each box?
    b) How many books left over?

7) 58 people make teams of 4 for a quiz.
    a) How many teams?
    b) The people left over read the quiz questions. How many people read the questions?

8) 576 bottles are filled on 4 machines. The same number of bottles are filled on each machine.
How many bottles does each machine fill?

9) 179 chickens are divided equally into 4 runs.
How many chickens are **not** in a run?

Dividing with remainders and calculation practice

Dividing by 6 with remainders

 I will learn to divide by 6 when there is a remainder.

Remember, remember

Dividing by 6 is the same as **sharing equally between 6** or **grouping in 6s**.

Sometimes, when you **divide or share**, there are some left over.

This is called a **remainder**.

Share 22 buttons between 6 children.

Each child gets 3 buttons.

There are 4 buttons left over.

There is a **remainder of 4**.

```
      3 r 4
   _____
 6 | 2 2
```

22 ÷ 6 = 3 r 4

Chapter 3 Whole numbers 1

Example

Work out 548 ÷ 6.

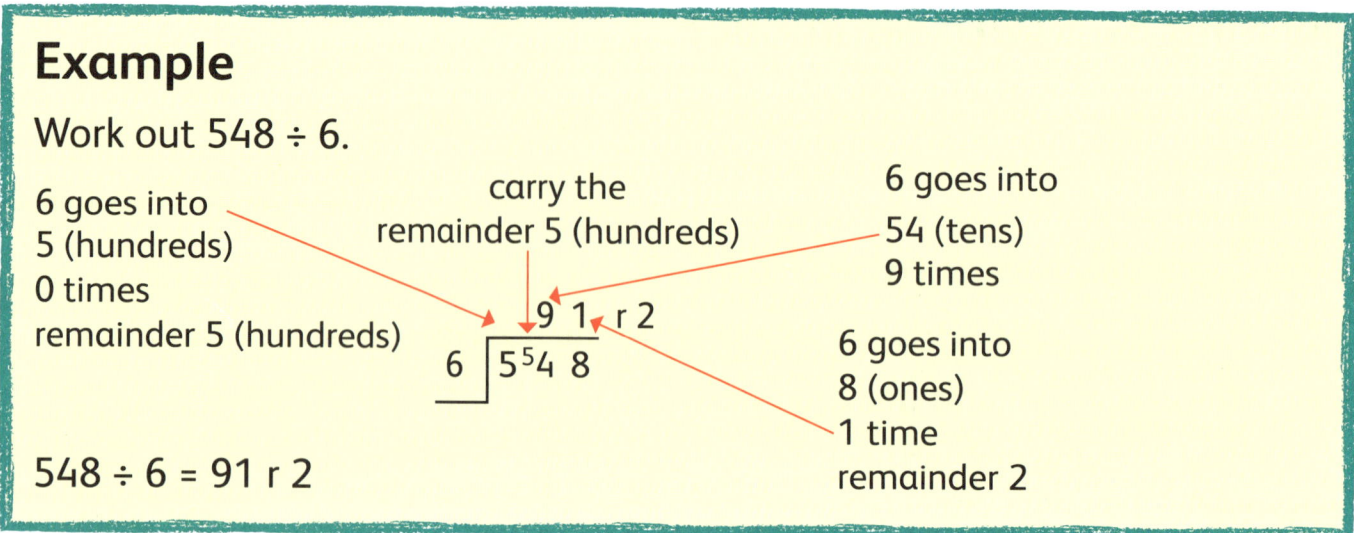

6 goes into 5 (hundreds) 0 times remainder 5 (hundreds)

carry the remainder 5 (hundreds)

6 goes into 54 (tens) 9 times

6 goes into 8 (ones) 1 time remainder 2

548 ÷ 6 = 91 r 2

Exercise 2

1) Knowing your 6 times table helps when dividing by 6.
   In your jotter, write the 6 times table up to 6 × 9.

2) Work out these divisions.
   You can use cubes or counters to help you.

   a) 7 ÷ 6

   b) 10 ÷ 6

3) Copy and complete:

   a) 6) 6 8 = □□ r □

   b) 6) 7 ¹8 = □□

   c) 6) 8 ²4 6 = □□□

   d) 6) 6 4 ⁴3 = □0□ r □

   e) 6) 7 ¹9 ¹2 = □□□

   f) 6) 4 ⁴7 ⁵5 = □□ r □

Dividing with remainders and calculation practice

4) Work out:
- a) 6⟌62
- b) 6⟌96
- c) 6⟌126
- d) 6⟌582
- e) 6⟌439
- f) 6⟌770

5) Copy and complete:

□□□□ r□
6⟌6 8 ²2 ⁴7

6) In your jotter, write each division using ⟌
Then work out the answer.
- a) 306 ÷ 6
- b) 858 ÷ 6
- c) 747 ÷ 6
- d) 9205 ÷ 6

7) 75 place napkins are tied in groups of 6.
- a) How many napkins are tied?
- b) How many napkins left over?

8) 128 cars are parked in rows of 6.
- a) How many full rows?
- b) How many cars left over?

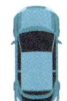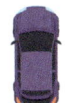

9) 1104 people are split between 6 hotels.
The same number of people stay at each hotel.
How many people at each hotel?

 10) 192 children are divided between 6 classes.
Each class is divided into 4 groups.
How many children in a group?

Chapter 3 Whole numbers 1

Dividing by 7 with remainders

 I will learn to divide by 7 when there is a remainder.

Remember, remember
Dividing by 7 is the same as **sharing equally between 7** or **grouping in 7s**.
Sometimes, when you **divide or share**, there are some left over.
This is called a **remainder**.

Share 15 counters between 7 children.

Each child gets 2 counters.
There is 1 counter left over.
There is a **remainder of 1**.

$$\begin{array}{r} 2\ \text{r}\ 1 \\ 7\overline{)15} \end{array}$$

$15 \div 7 = 2\ \text{r}\ 1$

Dividing with remainders and calculation practice

Example

Work out 789 ÷ 7.

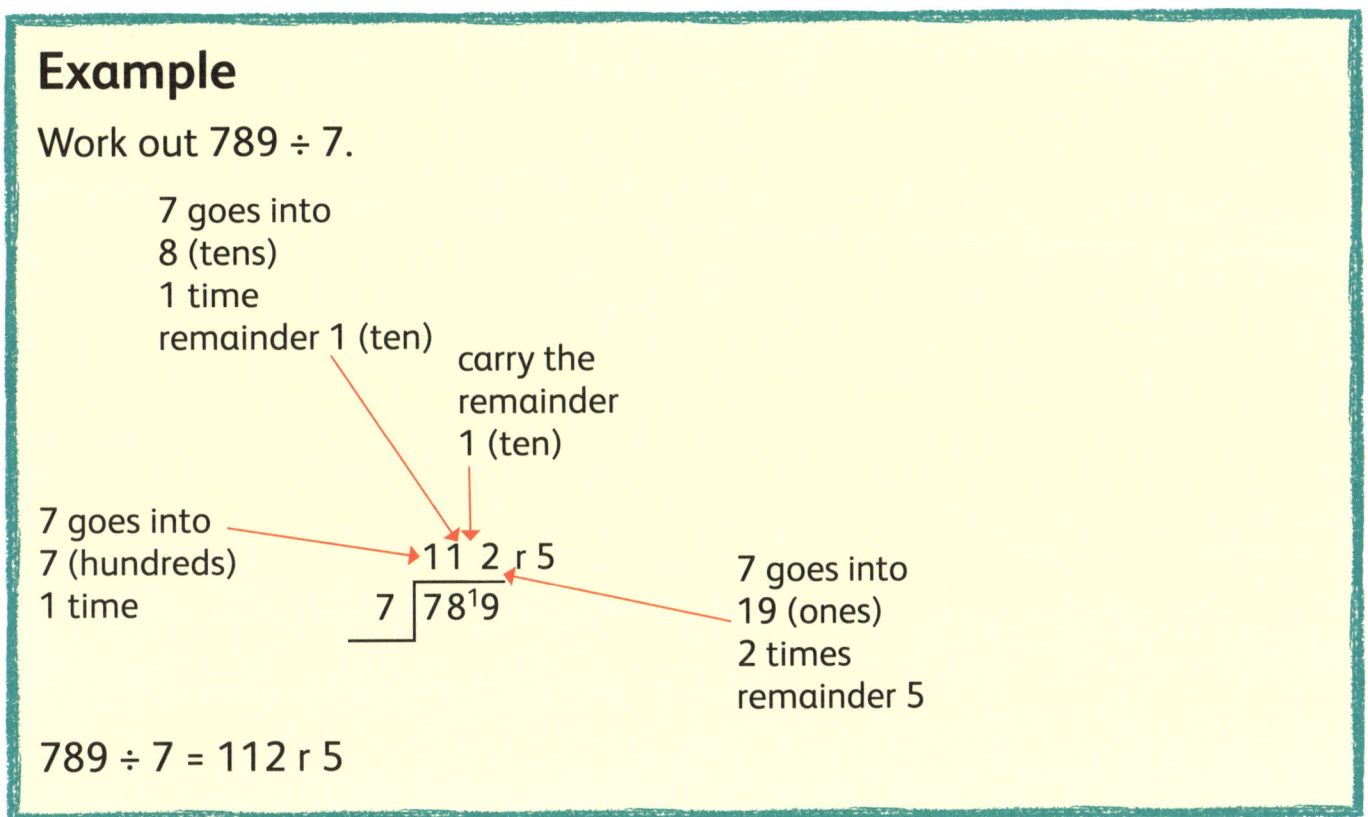

789 ÷ 7 = 112 r 5

Exercise 3

1) Knowing your 7 times table helps when dividing by 7.
   In your jotter, write the 7 times table up to 7 × 9.

2) Work out these divisions.
   You can use cubes or counters to help you.

   a) 9 ÷ 7

   b) 16 ÷ 7

Chapter 3 Whole numbers 1

3) Copy and complete:

a) 7) 7 5 = □□ r □

b) 7) 9 ²8 = □□

c) 7) 3 ³5 8 = □□ r □

d) 7) 9 ²2 ¹7 = □□□ r □

e) 7) 7 9 ²5 ⁴2 = □□□□

f) 7) 2 ²9 ¹4 6 = □□□ r □

4) Work out:

a) 7) 92

b) 7) 847

c) 7) 509

d) 7) 635

e) 7) 7844

f) 7) 9520

5) In your jotter, write each division using ⌐. Then work out the answer.

a) 73 ÷ 7
b) 882 ÷ 7
c) 948 ÷ 7
d) 8527 ÷ 7

6) 896 parcels are divided equally between 7 lorries. How many parcels on each lorry?

7) 160 police officers are divided equally between 7 stations. Any police officers left over are on duty in their car.
a) How many police officers in each station?
b) How many police officers on duty in their car?

8) How many weeks is 1001 days?

Dividing with remainders and calculation practice

Dividing by 8 with remainders

 I will learn to divide by 8 when there is a remainder.

Remember, remember

Dividing by 8 is the same as **sharing equally between 8** or **grouping in 8s**.
Sometimes, when you **divide or share**, there are some left over.
This is called a **remainder**.

Share 35 counters between 8 children.

Each child gets 4 counters.
There are 3 counters left over.
There is a **remainder of 3**.

```
    4 r 3
8 ) 3 5
```

35 ÷ 8 = 4 r 3

Chapter 3 Whole numbers 1

Example

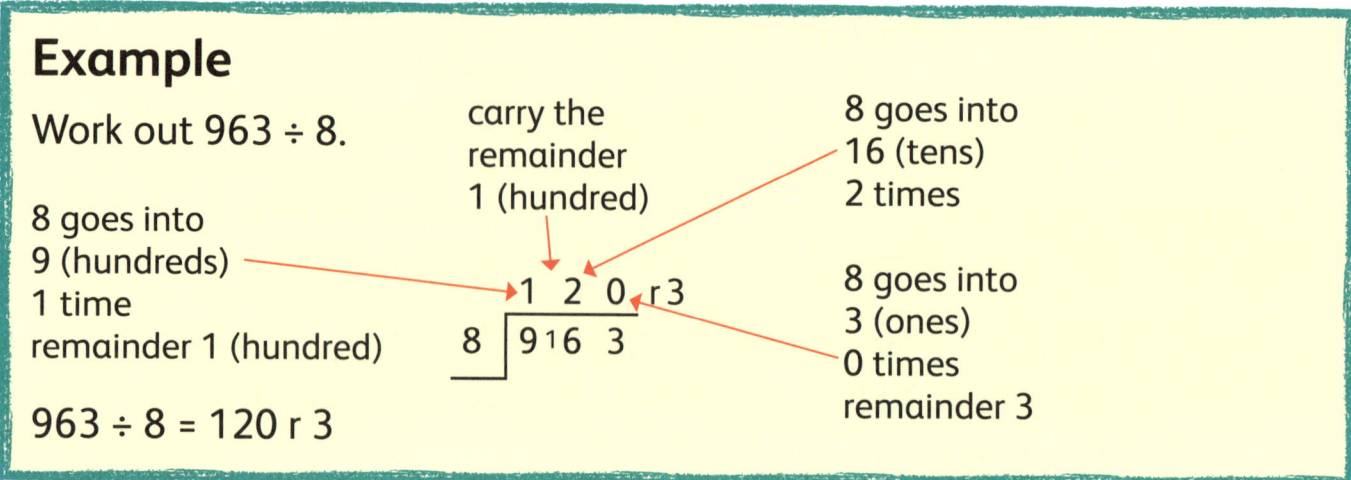

Work out 963 ÷ 8.

8 goes into 9 (hundreds) 1 time remainder 1 (hundred)

carry the remainder 1 (hundred)

8 goes into 16 (tens) 2 times

8 goes into 3 (ones) 0 times remainder 3

963 ÷ 8 = 120 r 3

Exercise 4

1) Knowing your 8 times table helps when dividing by 8.
   In your jotter, write the 8 times table up to 8 × 9.

2) Work out these divisions.
   You can use cubes or counters to help you.

   a) 9 ÷ 8
   b) 20 ÷ 8
   c) 27 ÷ 8

3) Copy and complete:

   a) 8) 8 6
   b) 8) 9 ¹6
   c) 8) 8 9 ¹3
   d) 8) 2 ²0 ⁴8
   e) 8) 5 ⁵7 ¹2
   f) 8) 9 ¹4 ⁶0 ⁴2

4) Work out:

   a) 8) 84
   b) 8) 99
   c) 8) 816
   d) 8) 928
   e) 8) 950
   f) 8) 2176
   g) 8) 3294
   h) 8) 4737

Dividing with remainders and calculation practice

5) In your jotter, write each division using ⌐
   Then work out the answer.
   a) 89 ÷ 8   b) 964 ÷ 8   c) 729 ÷ 8   d) 8253 ÷ 8

6) A train has 432 seats.
   There are 8 carriages.
   How many seats in each carriage?

7) A post office sells books of 8 stamps.
   The post office has 946 stamps
   for sale.

   a) How many complete books of stamps are for sale?
   b) How many stamps left over?

 8) The product of two numbers is 1464.
   One number is 8.
   What is the other number?

Dividing by 9 with remainders

 I will learn to divide by 9 when there is a remainder.

Remember, remember

Dividing by 9 is the same as **sharing equally between 9** or **grouping in 9s**.
Sometimes, when you **divide or share**, there are some left over.
This is called a **remainder**.

Chapter 3 Whole numbers 1

Share 25 counters between 9 children.

Each child gets 2 counters.

There are 7 counters left over.

There is a **remainder of 7**.

```
   2 r 7
9 ) 2 5
```

25 ÷ 9 = 2 r 7

Example

Work out 658 ÷ 9.

9 goes into
6 (hundreds)
0 times
remainder 6 (hundreds)

carry the remainder 6 (hundreds)

9 goes into 65 (tens)
7 times
remainder 2 (tens)

carry the remainder 2 (tens)

9 goes into 28 (ones)
3 times remainder 1

```
    7 3 r 1
9 ) 6 ⁶5 ²8
```

658 ÷ 9 = 73 r 1

Dividing with remainders and calculation practice

Exercise 5

1) Knowing your 9 times table helps when dividing by 9.
   In your jotter, write the 9 times table up to 9 × 9.

2) Work out these divisions.
   You can use cubes or counters to help you.
   a) 11 ÷ 9   b) 20 ÷ 9   c) 32 ÷ 9

3) Copy and complete:

   a) 9) 9 2 = □□ r □

   b) 9) 1 ¹0 ¹8 = □□

   c) 9) 3 ³0 ³6 = □□

   d) 9) 9 4 ⁴7 = □□□ r □

   e) 9) 9 6 ⁶4 = □□□ r □

   f) 9) 8 ⁸4 ³6 2 = □□□ r □

4) Work out:
   a) 9) 97   b) 9) 126   c) 9) 954   d) 9) 973
   e) 9) 282   f) 9) 5463   g) 9) 7852   h) 9) 5001

5) In your jotter, write each division using ⌐
   Then work out the answer.
   a) 10 ÷ 9   b) 100 ÷ 9   c) 1000 ÷ 9

6) Use your answers to question 5 to **predict** (make a good guess, without working out) the answer to 10 000 ÷ 9.

Chapter 3 Whole numbers 1

7) A sweet shop has 378 jars of sweets.
The owner puts 9 jars on each shelf.
How many shelves in the sweet shop?

8) 508 mugs are put in boxes.
Each box has 9 mugs.
   a) How many full boxes?
   b) How many mugs left over?

 9) A textbook contains 660 questions.
The questions are grouped into exercises.
There are 9 questions in every exercise.
The questions that are left over are challenge questions at the end of the textbook.
   a) How many exercises in the book?
   b) How many challenge questions in the book?

Mixed problems: more multiplication and division

 I will be able to solve problems involving × and ÷

Remember, remember

You should know how to **multiply** by 2, 3, 4, 5, 6, 7, 8 and 9.
You can multiply a 2-digit number by a 1-digit number in two ways.
You can use the grid method:

48 × 7 = 280 + 56
       = 336

×	4 tens	8 ones
7	28 tens = 280	56 ones = 56

Dividing with remainders and calculation practice

Or you can use a column method:

```
    Hundreds  Tens  Ones
                4     8
  ×             5     7
  _____
             3  3     6
```

You should also know how to **divide** by 2, 3, 4, 5, 6, 7, 8 and 9 **with** and **without remainders**.

Sometimes a problem will not tell you whether to multiply or divide. You must decide.

Exercise 6

1) Work out:
   a) 96 times 8
   b) 608 divided by 6
   c) 945 shared equally between 7
   d) 79 multiplied by 6
   e) 768 grouped in 8s.

2) Find the **missing** numbers.
   a) ___ ÷ 7 = 28
   b) ___ × 8 = 768
   c) ___ ÷ 6 = 98
   d) ___ × 9 = 558
   e) ___ × 4 = 1836
   f) ___ ÷ 8 = 448

These are either **multiplication or division** problems. You must decide which.

3) A group of 9 people hire bicycles.
   All bicycles are the same price to hire.
   The group pays £324.
   How much does each person pay?

Chapter 3 Whole numbers 1

4) Freddie is 12 years old.

Grandma is 6 times Freddie's age.

a) How old is Grandma?

Roman is $\frac{1}{8}$ of Grandma's age.

b) How old is Roman?

5) Daryna needs 140 cards.

They are sold in packs of 6.

How many packs should she buy?

6) Adam buys 165 roses.

He sells them in bunches of 6 roses.

a) How many full bunches does he make?

b) How many roses does he have left over?

Adam charges £9 for each bunch of roses.

He sells all his full bunches.

c) How much money does Adam take?

 7) There are 4 dog treats in a pack.

Sam has 2 dogs.

Each day a dog eats a whole pack.

How many treats do the dogs eat in 6 weeks?

Now try this!

Work with a partner to find the **missing** numbers.

1) $8 \overline{) 4 \square^1 \square}$ gives $5 \square$ r 1

2) $\square \overline{) 4\,4^\square\,{}^1 3\,{}^1 \square}$ gives $8 \square 3$

Challenge

Write your own missing number problem and swap problems with another pair.

Dividing with remainders and calculation practice

Revisit, review, revise

1) Copy and complete:

   a) 4) 4 5 = □□ r □

   b) 6) 8 ²4 = □□

   c) 7) 8 ¹4 9 = □□□ r □

   d) 4) 9 ¹4 ²7 = □□□ r □

   e) 9) 5 ⁵6 ²2 = □□ r □

   f) 8) 6 ⁶1 ⁵0 ²0 = □□□ r □

2) Work out:

   a) 967 ÷ 6   b) 725 ÷ 7   c) 1059 ÷ 4   d) 5032 ÷ 8

These are either multiplication or division problems.
You must decide which.

3) A baker makes 250 doughnuts.
   They are put in boxes of 4.

   a) How many full boxes?

   b) How many doughnuts left over?

4) Seven tickets to a festival cost £952.
   All tickets are the same price.
   How much is one ticket?

5) A transport company has 84 lorries.
   Each lorry has 6 wheels.
   How many wheels in total?

6) An artist uses stars in his work.
   Each star has 8 points.
   There are 872 points in the work.
   How many stars?

4 Sequences 1
Identifying patterns

Finding rules

💡 I will learn to explain and use a rule to find terms in a sequence.

A **sequence** is a list of numbers or objects in a given order.
Each number or object in a sequence is called a **term**.
Sometimes there is a rule connecting terms in a sequence.
The sequence: 1, 3, 5, 7, 9 … is the sequence of odd numbers.
To find the next term, 'add 2' to the previous term.
We can describe a sequence by giving the **first term** and the **term-to-term** rule.

Example

We can describe the sequence
3, 6, 12, 24 …
by writing:
first term = 3   term-to-term rule = ×2

Exercise 1

 1) In your jotter, write the next 2 terms in each sequence.
   a) 4, 8, 12, 16, 20 …
   b) 6, 12, 18, 24, 30 …
   c) 15, 20, 25, 30, 35 …
   d) 18, 27, 36, 45 …

Identifying patterns

2) Describe each sequence in question 1.
Copy and complete:
first term = _____ term-to-term rule = _____

3) In your jotter, write the next 2 terms in each sequence.
a) 5, 7, 9, 11 …
b) 10, 13, 16, 19 …
c) 16, 14, 12, 10 …
d) 3700, 3600, 3500, 3400 …
e) 52, 48, 44, 40 …
f) 64, 55, 46, 37 …
g) 1, 2, 4, 8 …
h) 800, 400, 200, 100 …

4) Describe each sequence in question 3.
Copy and complete:
first term = _____ term-to-term rule = _____

5) In your jotter, write the first 4 terms for each sequence.
a) first term = 10, term-to-term rule = add 1
b) first term = 30, term-to-term rule = subtract 3
c) first term = 5, term-to-term rule = multiply by 2
d) first term = 64, term-to-term rule = divide by 2
e) first term = 20, term-to-term rule = add $\frac{1}{2}$

6) a) In your jotter, write the next 2 terms in each sequence.
b) Describe each sequence giving the first term and the term-to-term rule.

i) $\frac{1}{4}, \frac{1}{2}, \frac{3}{4}, 1, 1\frac{1}{4}$ …
ii) 0.2, 0.4, 0.6, 0.8, 1 …
iii) 10, 9.5, 9, 8.5, 8 …
iv) $5, 4\frac{2}{3}, 4\frac{1}{3}, 4, 3\frac{2}{3}$ …

Chapter 4 Sequences 1

Special sequences

 I will learn to recognise sequences in the real world.

You will see sequences all around you.
The numbers of doors in a street may follow a sequence.

The **Fibonacci sequence** is a special sequence often found in nature.
It is named after a famous Italian mathematician.
The terms in a Fibonacci sequence are found by adding the previous two terms.

1, 1, 2, 3, 5, 8 …

The next term will be (5 + 8 =) 13
The Fibonacci sequence appears in nature in flowers and pine cones.
It even appears in space!

Exercise 2

 1) Here is a sequence made with sticks.

1 triangle   2 triangles   3 triangles

3 sticks   6 sticks   9 sticks

a) Draw the next term in the sequence.
b) The sequence for the number of sticks is: 3, 6, 9, 12 ....
Copy the sequence and write the next 3 terms.
c) Copy and complete:
Start with 3 sticks for 1 triangle and add on ____ sticks for each extra triangle.
d) How many sticks are needed for the **10th term**?

2) A baker makes cakes which are decorated with cherries.

1 cake (6 cherries)   2 cakes (12 cherries)   3 cakes (18 cherries)

a) How many cherries are needed for 4 cakes?
b) Copy the sequence 6, 12, 18 and write the next 3 terms.
c) Copy and complete:
Start with ____ cherries for 1 cake and add on ____ cherries for each extra cake.
d) How many cherries are needed for 9 cakes?

3) Here is another sequence made with sticks:

a) Draw the next term in the sequence.
b) How many sticks are needed for this term?

Chapter 4 Sequences 1

c) The pattern for the number of sticks needed is: 6, 11, 16 …
Copy the sequence and write the next 3 terms.

d) Copy the following and complete:
Start with ____ sticks for 1 house and add on ____ sticks for each extra house.

4) The Fibonacci sequence starts: 1, 1, 2, 3, 5, 8, 13 …
Continue the sequence to find the first 12 terms in the Fibonacci sequence.

5) Find the first 10 numbers in these Fibonacci sequences:
a) 3, 4, 7, 11 …
b) 2, 5, 7 …

6) Form your own Fibonacci sequence.
Start with any 2 numbers.
Find the first 10 terms in your sequence.

7) Here is a sequence made with blocks.

a) Draw the next term in the sequence.
b) The number of blocks in each term is: 1, 3, 6, 10.
Copy the sequence and write the next 3 terms.
c) The numbers in this sequence are called the **triangular numbers**.
Can you explain why?

Identifying patterns

> **Now try this!**
>
> Work in pairs.
>
> One player writes or draws the first five terms of a sequence.
>
> The second player has to spot the rule and write or draw the next term in the sequence. If they are correct, they score one point.
>
> Take it in turns.
>
> The first player to five points wins.

Revisit, review, revise

1) a) In your jotter, write the next two terms in each sequence.

   b) Describe each sequence:

   first term = _____

   term-to-term rule = _____.

   i) 15, 21, 27, 33, 39   ii) 60, 51, 42, 33
   iii) 1, 2, 4, 8 …   iv) 2, 16, 30, 44, 58 …
   v) 74, 69, 64, 59 …   vi) 9.8, 9.6, 9.4, 9.2 …

2) Here is a sequence made with bricks.

   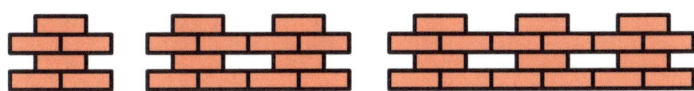

   a) How many bricks are needed for the **4th term**?

   b) Copy the sequence 6, 12, 18 and write the next 3 terms.

Chapter 4 Sequences 1

c) Copy and complete:

Start with ____ bricks for the first term and add on ____ bricks each time.

☀ 3) A Fibonacci sequence starts: 2, 2, 4, 6, 10 …

a) In your jotter, write the next 3 terms.

b) How many terms in the sequence are **smaller than 100**?

Multiples and factors
Finding multiples and factors

Multiples

> 💡 I will learn to find a multiple of a number.

Numbers in the 3 times table are called **multiples of 3**.
For example, the **first three (non-zero) multiples of 3** are **3, 6, 9** because:
3 × 1 = **3**   3 × 2 = **6**   3 × 3 = **9**

Numbers in the 4 times table are called **multiples of 4**.
For example, the **first three (non-zero) multiples of 4** are **4, 8, 12** because:
4 × 1 = **4**   4 × 2 = **8**   4 × 3 = **12**

Numbers in the 5 times table are called **multiples of 5**.
And so on.

Zero (0) is a multiple of every number, because it is in every times table.
For example, 3 × 0 = 0   4 × 0 = 0

Exercise 1

 1) Copy and complete the first six (non-zero) multiples of 5:
5, 10, 15, ___, ___, ___

2) In your jotter, write the first five (non-zero) multiples of:
   a) 3   b) 4   c) 2   d) 6   e) 10

Chapter 5 Multiples and factors

3) In your jotter, write **true** or **false** for each statement.
   a) 40 is a multiple of 10
   b) 30 is a multiple of 4
   c) 50 is a multiple of 5
   d) 45 is a multiple of 7
   e) 42 is a multiple of 6
   f) 56 is a multiple of 9

4) In your jotter, write the first ten (non-zero) multiples of 8.

5) In your jotter, write **all**:
   a) the multiples of 3 between 23 and 32
   b) the multiples of 6 between 20 and 40
   c) the multiples of 7 that are less than 25
   d) the multiples of 9 that are less than 50

6) Mia says:

   I am thinking of a number.
   It is less than 20 (but not zero).
   My number is a multiple of 2 and 3.

   In your jotter, write **all** possibilities for Mia's number.

 7) Sam says:

   I have a coin which is less than £1.
   The coin is a multiple of 5.
   The tens digit of the coin is even.
   What is my coin?

Finding multiples and factors

Factors

 I will learn to find a factor of a number.

Numbers that divide exactly into 6 (with no remainder) are called **factors of 6**.
For example, 1, 2, 3 and 6 are factors of 6 because
6 ÷ **1** = 6   6 ÷ **2** = 3   6 ÷ **3** = 2   6 ÷ **6** = 1
and there are no remainders.

Numbers that divide exactly into 10 (with no remainder) are called **factors of 10**.
For example, 1, 2, 5 and 10 are factors of 10 because
10 ÷ **1** = 10   10 ÷ **2** = 5   10 ÷ **5** = 2   10 ÷ **10** = 1
and there are no remainders.
And so on.

Factors can usually be written in pairs.
Factor pairs of **6** are:
1 and 6 (because 1 × 6 = 6); 2 and 3 (because 2 × 3 = 6).
Factor pairs of **10** are:
1 and 10 (because 1 × 10 = 10); 2 and 5 (because 2 × 5 = 10).

Exercise 2

1) In your jotter, write a factor of 4.

2) In your jotter, write all four factors of 8.

Chapter 5 Multiples and factors

3) Copy and complete to find all the factor pairs of 12:
   a) 1 × ___ = 12   1 and ___ are factors of 12
   b) 2 × ___ = 12   2 and ___ are factors of 12
   c) 3 × ___ = 12   3 and ___ are factors of 12

4) In your jotter, write the:
   a) two factors of 11   b) five factors of 16

5) In your jotter, write **true** or **false** for each statement.
   a) 4 is a factor of 32   b) 9 is a factor of 45
   c) 6 is a factor of 34   d) 5 is a factor of 35
   e) 7 is a factor of 48   f) 56 is a factor of 8

6) In your jotter, write all the factors of:
   a) 7   b) 9   c) 15

7) In your jotter, write two even numbers that are a factor pair of 20.

8) Billy has £24.
   Can he share this money equally between:
   a) 2 pots   b) 3 pots   c) 4 pots
   d) 5 pots   e) 6 pots   f) 7 pots
   g) 8 pots?
   Answer **yes** or **no** for each.

Finding multiples and factors

Now try this!

Work in pairs.

There are 36 children in a singing group.

They must stand in rows on the stage.

Each row must have an equal number of children.

Find all the different ways they can stand in rows on the stage.

Revisit, review, revise

1) In your jotter, write the first six (non-zero) multiples of:
   a) 2
   b) 10
   c) 9
   d) 7

2) In your jotter, write all the multiples of 5 between 26 and 46.

3) In your jotter, write all the factors of 25.

4) Copy and complete to find all the factor pairs of 30:
   a) 1 × ____ = 30   1 and ____ are factors of 30
   b) 2 × ____ = 30   2 and ____ are factors of 30
   c) 3 × ____ = 30   3 and ____ are factors of 30
   d) 5 × ____ = 30   5 and ____ are factors of 30.

5) Here are some numbers:

10	8	24	16	20

   a) Which are multiples of 4?
   b) Which are factors of 20?
   c) Which is **both** a multiple of 4 and a factor of 20?

6 Time
Working with units of time

Units of time

 I will learn the common units of time.

There are many different units of time:
years, months, weeks, days, hours, minutes, seconds.

Remember, remember
There are 12 **months** in 1 **year**.
There are 365 **days** in 1 **year** (or 366 days in a leap year).
There are 24 **hours** in 1 **day**.
There are 60 **minutes** in 1 **hour**.
There are 60 **seconds** in 1 **minute**.

Exercise 1

1) How many seconds in:
   a) 4 minutes
   b) $1\frac{1}{2}$ minutes
   c) 6 minutes
   d) three and a half minutes?

2) How many minutes in:
   a) 4 hours
   b) $2\frac{1}{2}$ hours
   c) $1\frac{1}{4}$ hours
   d) five and a half hours?

3) How many months in:
   a) 2 years
   b) 10 years
   c) $1\frac{1}{2}$ years?

Working with units of time

4) Choose the unit of time from the box that you would use to measure:
   a) how long you can hold your breath
   b) the length of a maths lesson
   c) the time until your birthday
   d) your age
   e) how long you sleep at night
   f) how long the school term is
   g) how long until the weekend.

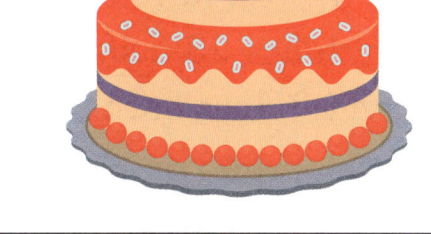

| seconds | minutes | hours | days | weeks | months | years |

5) How many days from 25th November to 7th December, including both dates?

| NOVEMBER 2023 |
S	M	T	W	T	F	S
			1	2	3	4
5	6	7	8	9	10	11
12	13	14	15	16	17	18
19	20	21	22	23	24	(25)
26	27	28	29	30		

| DECEMBER 2023 |
S	M	T	W	T	F	S
					1	2
3	4	5	6	(7)	8	9
10	11	12	13	14	15	16
17	18	19	20	21	22	23
24	25	26	27	28	29	30
31						

6) Put these lengths of time in order from shortest to longest:

| 2 days | $\frac{1}{4}$ hour | 40 minutes | January | 1 year |
| 1000 days | 7 seconds | $\frac{1}{2}$ minute | | |

Years, months and dates

 I will learn to interpret and use a calendar and write dates.

There are 365 days in a year (and 366 in a leap year).
Leap years occur every 4 years.
These are leap years: 2020, 2024, 2028, 2032 … etc.

Chapter 6 Time

Remember, remember

30 days have September,
April, June and November.
All the rest have 31, except February,
which has 28 days clear
and 29 days in each leap year.

A calendar can help us to keep track of time. This calendar shows the year 2024.

2024

JANUARY						
S	M	T	W	T	F	S
	1	2	3	4	5	6
7	8	9	10	11	12	13
14	15	16	17	18	19	20
21	22	23	24	25	26	27
28	29	30	31			

FEBRUARY						
S	M	T	W	T	F	S
				1	2	3
4	5	6	7	8	9	10
11	12	13	14	15	16	17
18	19	20	21	22	23	24
25	26	27	28	29		

MARCH						
S	M	T	W	T	F	S
					1	2
3	4	5	6	7	8	9
10	11	12	13	14	15	16
17	18	19	20	21	22	23
24	25	26	27	28	29	30
31						

APRIL						
S	M	T	W	T	F	S
	1	2	3	4	5	6
7	8	9	10	11	12	13
14	15	16	17	18	19	20
21	22	23	24	25	26	27
28	29	30				

MAY						
S	M	T	W	T	F	S
			1	2	3	4
5	6	7	8	9	10	11
12	13	14	15	16	17	18
19	20	21	22	23	24	25
26	27	28	29	30	31	

JUNE						
S	M	T	W	T	F	S
						1
2	3	4	5	6	7	8
9	10	11	12	13	14	15
16	17	18	19	20	21	22
23	24	25	26	27	28	29
30						

JULY						
S	M	T	W	T	F	S
	1	2	3	4	5	6
7	8	9	10	11	12	13
14	15	16	17	18	19	20
21	22	23	24	25	26	27
28	29	30	31			

AUGUST						
S	M	T	W	T	F	S
				1	2	3
4	5	6	7	8	9	10
11	12	13	14	15	16	17
18	19	20	21	22	23	24
25	26	27	28	29	30	31

SEPTEMBER						
S	M	T	W	T	F	S
1	2	3	4	5	6	7
8	9	10	11	12	13	14
15	16	17	18	19	20	21
22	23	24	25	26	27	28
29	30					

OCTOBER						
S	M	T	W	T	F	S
		1	2	3	4	5
6	7	8	9	10	11	12
13	14	15	16	17	18	19
20	21	22	23	24	25	26
27	28	29	30	31		

NOVEMBER						
S	M	T	W	T	F	S
					1	2
3	4	5	6	7	8	9
10	11	12	13	14	15	16
17	18	19	20	21	22	23
24	25	26	27	28	29	30

DECEMBER						
S	M	T	W	T	F	S
1	2	3	4	5	6	7
8	9	10	11	12	13	14
15	16	17	18	19	20	21
22	23	24	25	26	27	28
29	30	31				

Working with units of time

Exercise 2

1) a) What is the **first** month of the year?
   b) What is the **last** month of the year?
   c) Which month comes just **after** July?
   d) Which month comes just **before** May?

2) How many days in:
   a) January
   b) March
   c) April
   d) June
   e) August
   f) October
   g) November
   h) December?

3) What is the:
   a) 6th month
   b) 3rd month
   c) 10th month
   d) 8th month?

Remember, remember

The date 3rd January 2015 can be written in different ways:

3rd Jan 2015   03/01/15   03.01.15

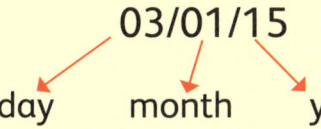

day   month   year   day   month   year

4) In your jotter, write each date using six digits.
   a) 23rd February 2014
   b) 19th April 2003
   c) 22nd July 2014
   d) 18th August 1997
   e) 7th June 1985
   f) 2nd February 2020

Chapter 6 Time

5) In your jotter, write each date in words.

For example: 1/1/27 would be 1st January 2027.

a) 19/5/28　　　b) 5/9/24　　　c) 30/11/20

6) Here is Nadim's calendar:

JANUARY 2025						
MON	TUE	WED	THU	FRI	SAT	SUN
		1 New Year's Day	2	3 Golf	4	5 Tennis
6 Back to school	7	8	9	10	11 Swimming	12
13	14	15 Rob's birthday	16	17	18	19
20	21	22	23	24	25 Burns night	26
27	28	29	30	31 Meet Ali		

a) What is Nadim doing on 6th January 2025?
b) What day is 23/1/25?
c) How many days **after** Rob's birthday is Burns night?
d) What date does Nadim play tennis?
e) Two days before Nadim goes swimming, he buys goggles. What date is this?

 7) Which years are leap years?

a) 2024　　　b) 2026　　　c) 2030
d) 2036　　　e) 2040　　　f) 2042

 8) Which of these dates will **never** happen?

a) 29th February 2028　　　b) 31st September 2025
c) 31st July 2080　　　d) 28.2.25
e) 31.6.30　　　f) 32/5/32

Working with units of time

Hours, minutes and seconds

 I will learn to convert between units of time and to add and subtract units of time.

Time can be written in different units.
1 minute = 60 seconds   1 hour = 60 minutes
2 minutes = 2 × 60 = 120 seconds   2 hours = 2 × 60 = 120 minutes
3 minutes = 3 × 60 = 180 seconds   3 hours = 3 × 60 = 180 minutes
You can convert between units of time.

Example

100 seconds = 60 seconds + 40 seconds
            = 1 minute and 40 seconds

200 minutes = 180 minutes + 20 minutes
            = 3 hours and 20 minutes

Exercise 3

 1) Continue the pattern up to 12 minutes.
- 1 minute = 60 seconds
- 2 minutes = 120 seconds
- 3 minutes = 180 seconds ...

 2) How many minutes in:
a) 1 hour
b) 10 hours
c) 4 hours
d) $\frac{1}{2}$ hour
e) $\frac{1}{4}$ hour
f) $\frac{3}{4}$ hour?

Chapter 6 Time

3) Change these into minutes and seconds:
   a) 90 seconds
   b) 148 seconds
   c) 220 seconds
   d) 620 seconds
   e) 172 seconds
   f) 335 seconds.

4) Change these into hours and minutes:
   a) 85 minutes
   b) 125 minutes
   c) 245 minutes
   d) 700 minutes
   e) 550 minutes
   f) 453 minutes.

To solve problems involving hours, minutes and seconds, work separately with each unit before combining them.

Example

Ami walks for 1 hour 35 minutes. She then rests before walking for another 2 hours 42 minutes.

How long does she walk **altogether**?

Add together the hours: 1 + 2 = 3 hours

Add together the minutes: 35 + 42 = 77 minutes = 1 hour 17 minutes

Total time = 3 hours + 1 hour 17 minutes = 4 hours 17 minutes

5) Work out:
   a) 3 minutes 5 seconds + 7 minutes 29 seconds
   b) 11 hours 45 minutes + 2 hours 20 minutes
   c) 5 minutes 35 seconds + 1 minute 30 seconds
   d) 1 hour 45 minutes + 42 minutes
   e) 4 minutes 18 seconds + 2 minutes 53 seconds.

6) An online video lasts 3 minutes 12 seconds.
   Erin watches it **twice**.
   How long does this take?

Working with units of time

 7) Amir plays a video game for 1 hour 35 minutes on Monday and 1 hour 43 minutes on Tuesday.
How long is this **altogether**?

Now try this!

Look at your school timetable.
Work out the total length of your maths lessons over a whole week.
Work out the total length of your playtimes over a week.
Work out the total length of time you will spend in maths lessons this term.

Timetables

 I will learn to read a timetable.

Here is a train timetable:

A train leaves from Shawfair at 07:06.

The 08:54 from Edinburgh arrives at Gorebridge at 09:21.

Edinburgh	0543	0622	0651	0724	0753	0824	0854	0924	0954
Brunstane	0551	0630	0659	0731	0800	0831	0902	0931	1001
Newcraighall	0556	0634	0703	0735	0804	0835	0906	0935	1005
Shawfair	0602	0637	0706	0738	0807	0838	0909	0938	1008
Eskbank	0608	0642	0711	0743	0812	0843	0913	0943	1013
Newtongrange	0612	0645	0714	0746	0815	0846	0916	0946	1016
Gorebridge	0621	0650	0719	0751	0820	0851	0921	0951	1021
Stow	-	0708	0737	0808	0837	0908	-	1008	-
Galashiels	0645	0716	0746	0817	0846	0917	0945	1017	1045
Tweedbank	0649	0721	0750	0821	0850	0921	0950	1021	1049

This train does not stop at Stow.

These are the different times a train leaves Eskbank.

Chapter 6 Time

Use a number line to work out how long the 08:54 train from Edinburgh takes to get to Gorebridge.

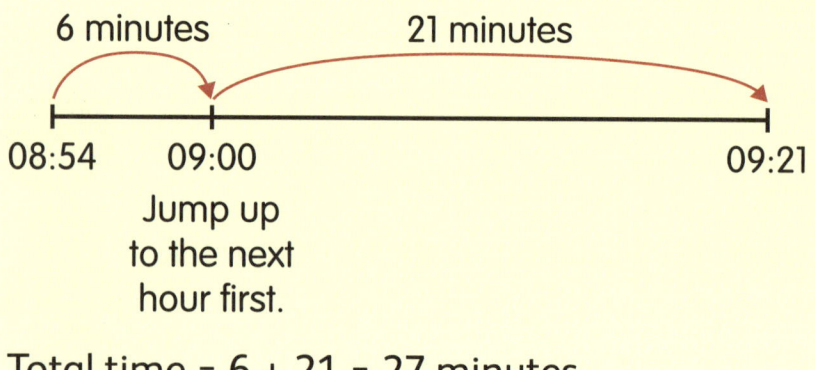

Jump up to the next hour first.

Total time = 6 + 21 = 27 minutes

Exercise 4

 1) Here is part of a train timetable from Edinburgh to Eskbank:

Edinburgh	09:32
Brunstane	09:37
Newcraighall	09:41
Shawfair	09:47
Eskbank	09:54

a) What time does the train leave from Brunstane?

b) What time is the train in Shawfair?

c) How long is the journey from Edinburgh to Eskbank?

2) Here is the film timetable for a cinema:

	Monday	Tuesday	Wednesday	Thursday
Skipper	11:25	11:00	10:45	-
Fantasy Feet	15:30	17:15	12:25	18:45
Cranky Clown	11:35	09:30	-	14:25

Working with units of time

a) What time can you watch *Fantasy Feet* on Wednesday?
b) Which day is *Skipper* not showing?
c) Which day can you watch *Cranky Clown* before 11:00?
d) *Fantasy Feet* lasts 1 hour and 37 minutes.
What time does *Fantasy Feet* end on:
i) Monday
ii) Tuesday
iii) Wednesday
iv) Thursday?
Draw number lines to help.

 3) A Sunday train service timetable is shown.

Balloch	11:15	13:05	15:20	18:00	22:05
Partick	11:47	13:37	15:52	-----	22:37
Airdrie	12:17	-----	16:22	-----	23:09
Bathgate	12:32	14:22	16:37	-----	23:54
Edinburgh	12:59	14:39	16:54	19:00	00:11

a) When does the 15:20 from Balloch arrive in Edinburgh?
b) When does the 23:09 from Airdrie arrive in Bathgate?
c) How long does the 12:32 Bathgate train take to reach Edinburgh?
d) I arrive in Airdrie at 4:15 p.m.
How long do I have to wait for a train to Edinburgh?
e) It takes me 22 minutes to walk from my house in Airdrie to the station.
If I catch the 12:17 to Edinburgh, what is the latest time I can leave home?

Now try this!

Find a timetable online, or use one which your teacher has.
In your jotter, write five questions for a partner to answer.
Work out the answers, then mark your partner's work.

Chapter 6 Time

Longer time intervals

💡 I will learn to calculate a time interval.

Remember, remember
There are 24 hours in a day.
00:00 is midnight; it comes after 23:59.

You can use a number line to help work out a time interval.

Example 1
A late film starts at 22:50 and is 2 hours 15 minutes long. When does it end?

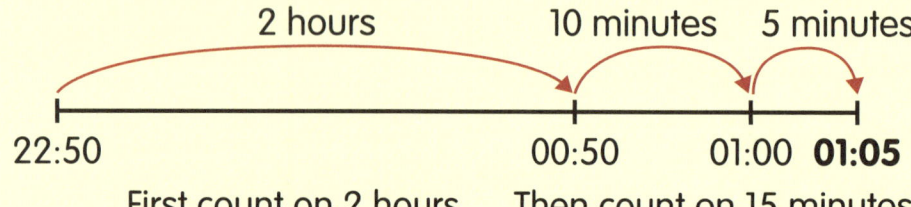

First count on 2 hours Then count on 15 minutes; it helps to break this down into 10 minutes and 5 minutes

Example 2
Erin goes to sleep at 21:15. She wakes up at 07:30 the next morning. How long does she sleep?

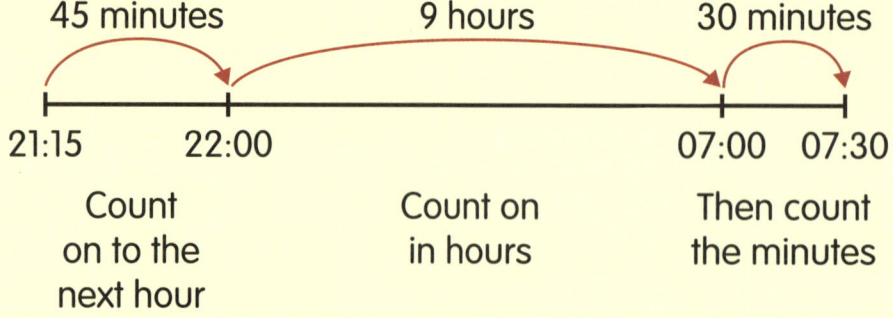

Count on to the next hour   Count on in hours   Then count the minutes

Add together the minutes:
45 minutes + 30 minutes = 75 minutes = 1 hour 15 minutes

Add together the hours and minutes:
9 hours + 1 hour 15 minutes = 10 hours 15 minutes

Working with units of time

Exercise 5

1) How long is it from 3:30 p.m. to 8:15 p.m?

   Copy and complete the number line:

   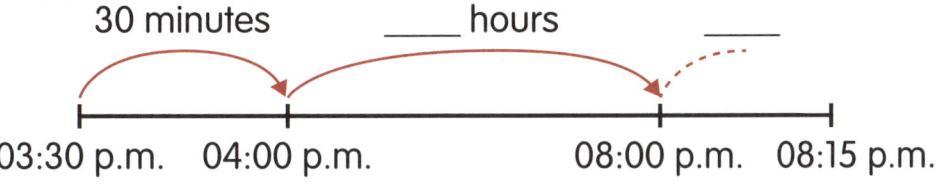

2) How long is it from 8:30 a.m. to 3:15 p.m.?
   Draw a number line to help.

3) How long is it from:
   a) 4:15 p.m. to 10:40 p.m.
   b) midnight to 2:52 a.m.
   c) 8:45 a.m. to 11:20 a.m.
   d) 17:35 to 21:30
   e) 13:15 to 21:05
   f) 9 a.m. to 8:10 p.m.?

4) These clocks show the start and end time of a festival.
   How long is the festival?

   Start a.m.   End p.m.

5) Viktor leaves home at 06:45.
   It takes him 1 hour 18 minutes to get to work.
   What time does he arrive?

6) The journey from Inverness to Paris takes 10 hours 10 minutes.
   Sammi leaves Inverness at 11:25 p.m. on Saturday.
   What time does she arrive in Paris on Sunday?

Chapter 6 Time

7) This is a departures board at an airport.

a) The flight to Paris takes 1 hour 55 minutes.
   What time does it arrive in Paris?

b) The flight to Atlanta arrives at 23:25.
   How long does it take?

c) The flight to Frankfurt takes 1 hour 35 minutes.
   It is delayed by 20 minutes.
   When does it actually arrive?

Revisit, review, revise

1) a) How many seconds in 5 minutes?
   b) How many months in a year?
   c) How many days in a leap year?

2) What unit of time would you use to measure how long:
   a) it takes to eat a sandwich
   b) it takes to say the alphabet
   c) it takes to fly to America?

3) How many months of the year have exactly 30 days?

4) In your jotter, write 29th June 2025 using 6 digits.

Working with units of time

5) Janelle runs a race in 12 minutes and 18 seconds.

   Jade runs it in 11 minutes and 37 seconds.

   Who wins and by how many seconds?

6) Marie records how long she spends on her homework each night:

Monday	35 minutes
Tuesday	20 minutes
Wednesday	0 minutes
Thursday	15 minutes
Friday	45 minutes

How long does she spend in **total**?

Give the answer in hours and minutes.

7) Here is a cruise timetable.

	Ullapool	Mallaig	Fort William	Oban	Greenock	Ayr	Stranraer
Early	08:45	11:34	15:20	17:25	19:58	21:12	22:45
Late	13:30	16:19	21:05	- - -	- - -	- - -	- - -

a) How long does it take from Mallaig to Fort William?

b) How long does the Early cruise take from Ullapool to Stranraer?

c) If both cruises take the same time, when will the Late cruise be in Stranraer?

7 Decimal fractions 1
Decimals to 1, 2 and 3 decimal places

Decimal fractions: thousandths

 I will learn to read, interpret and write decimal thousandths.

Remember, remember

You should already know about thousands, hundreds, tens, ones, tenths and hundredths and that the **place of a digit** tells you the **size of the number**.

Thousands	Hundreds	Tens	Ones	tenths	hundreds
3	6	2	9	1	7

= 3 thousand, 6 hundred and twenty-nine point one seven

Here is a square:

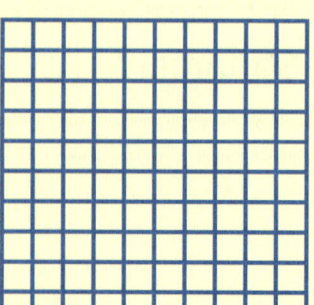

It is split into 100 equal parts.
One part represents:

$\frac{1}{100}$ = 0.01 (1 hundredth).

Decimals to 1, 2 and 3 decimal places

Now imagine each small square is split into 10 equal parts.
The grid is now split into 1000 equal parts.
One part represents $\frac{1}{1000}$ = 0.001 (1 thousandth).

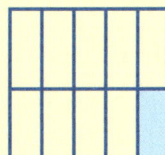

Ones	.	tenths	hundredths	thousandths
0	.	0	0	1

Exercise 1

 1) Copy and complete:

Ones	.	tenths	hundredths	thousandths
2	.	0	3	7

= 2. ___ ___ ___
= 2 ones, 0 tenths, ___ hundredths and 7 ___

2) Copy and complete:

a) 0.003 = 3 ___

b) 0.014 = 1 hundredth and ___ thousandths

c) 9.825 = 9 ones, 8 ___, 2 ___ and 5 ___

d) 12.416 = 1 ___, 2 ___, ___ tenths, ___ hundredths and ___ thousandths.

3) In the decimal **26.395**, what does the:

a) 2 represent   b) 9 represent   c) 5 represent?

4) What does the **4** represent in each number?

a) 0.014   b) 0.041   c) 4.015
d) 410.51   e) 4055.1   f) 5000.114

Chapter 7 Decimal fractions 1

5) Mentally work out what number is:
   a) 0.001 more than 6.005
   b) 0.003 more than 8.274
   c) 0.006 more than 471.052
   d) 0.002 less than 46.703

When ordering numbers, look at the digit with the **largest** place value first. Then look at the digit with the **next largest** place value and so on.

Example 1
When ordering 471.5, 184.2 and 612.1, look at the hundreds first.
- 612.1 has the **most hundreds**; 612.1 is the **biggest**.
- 184.2 has the **fewest hundreds**; 184.2 is the **smallest**.

The numbers in **order of size**, starting with the **biggest** are:
612.1, 471.5, 184.2

Example 2
When ordering 5.61, 1.95 and 5.7, look at the ones first, then the tenths.
- 5.61 and 5.7 have the most ones. One of these numbers is the biggest.
- 5.61 has 6 tenths and 5.7 has 7 tenths.
- 5.7 has **more tenths**; 5.7 is the **biggest**.

The numbers in **order of size**, starting with the **biggest** are: 5.7, 5.61, 1.95

Example 3
When ordering 0.704, 1.02 and 0.73, look at the ones first, then the tenths, then the hundredths.
- 1.02 has the **most ones**; 1.02 is the **biggest**.
- 0.704 and 0.73 both have 0 ones and 7 tenths.
- 0.704 has 0 hundredths and 0.73 has 3 hundredths.
- 0.73 has **more hundredths**; 0.73 is **bigger** than 0.704

The numbers in **order of size**, starting with the **biggest** are:
1.02, 0.73, 0.704

Decimals to 1, 2 and 3 decimal places

6) Here are some numbers: 8.36, 8.34, 8.307
   They all have 8 ones and 3 tenths.
   a) The number with the most hundredths is the biggest.
      Which number is the **biggest**?
   b) The number with the fewest hundredths is the smallest.
      Which number is the **smallest**?
   c) In your jotter, write the numbers in order of size, starting with the **biggest**.

7) In your jotter, write the **smallest** number in each group.
   a) 9.3, 9.1, 9.8
   b) 65.4, 65.3, 65.04
   c) 3.67, 3.02, 3.405
   d) 0.051, 0.063, 0.07
   e) 0.006, 0.016, 0.06
   f) 7.322, 7.5, 7.324

 8) In your jotter, write each group of numbers in question 8 in order of size. Start with the **smallest**.

Rounding to 1 decimal place

 I will learn to round a decimal to 1 decimal place.

Remember, remember

Here is a number line:

5 — 5.7 — 6

The arrow points to 5.7

Chapter 7 Decimal fractions 1

The **last whole number** before 5.7 is 5
The **next whole number** after 5.7 is 6
5.7 is **closer** to 6 than to 5
We say that '5.7, **rounded to the nearest whole number**, is 6'
When we round to the **last whole number**, it is called **rounding down**.
When we round to the **next whole number**, it is called **rounding up**.
Sometimes an arrow points **exactly half way** between two numbers.
When a number is **half way**, we **round up**.

A number with **one digit after the decimal point (tenths)** is written to **1 decimal place**.

Here, the arrow points to 3.17
3.17 lies between 3.1 and 3.2
3.17 is **closer** to 3.2 than to 3.1
We say that '3.17, **rounded to 1 decimal place**, is 3.2'

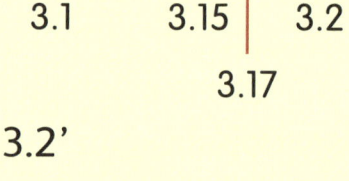

Here, the arrow points to 19.83
19.83 lies between 19.8 and 19.9
19.83 is **closer** to 19.8 than to 19.9
We say that '19.83, **rounded to 1 decimal place**, is 19.8'

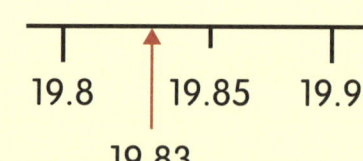

Decimals to 1, 2 and 3 decimal places

Exercise 2

 1) Look at the number line.

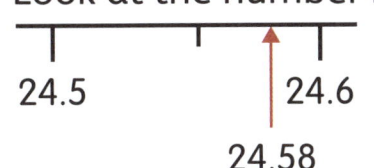

In your jotter, write the **missing** numbers.
a) 24.58 lies between ____ and ____
b) 24.58 is closer to ____ than to ____
c) 24.58 rounds to ____ (to 1 decimal place).

2 a) Sketch a number line from 4.6 to 4.7
b) Mark **4.62** on your number line.
c) Does 4.62 round to 4.6 or 4.7 (to 1 decimal place)?

3) a) Sketch a number line from 37.2 to 37.3
b) Mark **37.26** on your number line.
c) Does 37.26 round to 37.2 or 37.3 (to 1 decimal place)?

4) a) Imagine a number line between 10.5 and 10.6
b) Imagine **10.59** on your number line.
c) Does 10.59 round to 10.5 or 10.6 (to 1 decimal place)?

5) a) Imagine a number line between 22.3 and 22.4
b) Imagine **22.31** on your number line.
c) Does 22.31 round to 22.3 or 22.4 (to 1 decimal place)?

Chapter 7 Decimal fractions 1

6) a) Copy the numbers you have worked with so far.

| 3.1⑦ | 19.83 | 24.58 | 4.62 | 37.26 | 10.59 | 22.31 |

Circle the **second digit after the decimal point** in each number. The first one is done for you.

b) Below each number, write it rounded to 1 decimal place. Use the Examples and your answers to questions 1–5 to help you.

c) Zara says:

> When rounding to 1 decimal place, the **second digit after the decimal point** tells you whether to round up or round down.
> If it is **0**, **1**, **2**, **3** or **4**, then **round down**.
> If it is **5**, **6**, **7**, **8** or **9**, then **round up**.

Look at the numbers you have rounded so far.
Is Zara correct?
Yes or no.

7) Choose the number in the bracket that each decimal rounds to when rounded to 1 decimal place.
   a) 3.46 (3.4 or 3.5)
   b) 4.72 (4.7 or 4.8)
   c) 1.07 (1.0 or 1.1)
   d) 12.75 (12.7 or 12.8)
   e) 9.96 (9.9 or 10.0)
   f) 13.03 (13.0 or 13.1)
   g) 0.07 (0.0 or 0.1)
   h) 4.99 (4.9 or 5.0)

8) Copy and complete each pair of statements:
   a) 6.48 lies between 6.4 and 6.5   6.48 rounds to ___
                                         (to 1 decimal place)
   b) 13.81 lies between 13.8 and ___   13.81 rounds to ___
                                         (to 1 decimal place)

Decimals to 1, 2 and 3 decimal places

c) 1.78 lies between ____ and ____   1.78 rounds to ____
(to 1 decimal place)

d) 7.95 lies between ____ and 8.0   7.95 rounds to ____
(to 1 decimal place)

9) Round these numbers to 1 decimal place.
   a) 7.84   b) 92.87   c) 101.72   d) 19.95

 10) A footpath is 1.66 km long.

Is it more accurate to round the length to the **nearest km** or **1 decimal place**?

Rounding to 2 decimal places

💡 I will learn to round a decimal to 2 decimal places.

A number with **two digits after the decimal point** (tenths and hundredths) is written to **2 decimal places**.

Here, the arrow points to 8.126
8.126 lies between 8.12 and 8.13
8.126 is **closer** to 8.13 than to 8.12
We say that '8.126, **rounded to 2 decimal places**, is 8.13'

Here, the arrow points to 35.972
35.972 lies between 35.97 and 35.98
35.972 is **closer** to 35.97 than to 35.98
We say that '35.972, **rounded to 2 decimal places**, is 35.97'

When rounding to **2 decimal places**, you can look at the **third digit after the decimal point**.
- If it is **0, 1, 2, 3** or **4**, then **round down**.
- If it is **5, 6, 7, 8** or **9**, then **round up**.

89

Chapter 7 Decimal fractions 1

Exercise 3

1) Look at the number line.

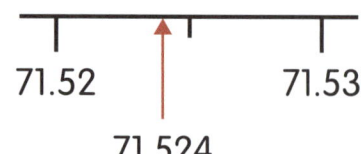

In your jotter, write the **missing** numbers.
a) 71.524 lies between ____ and ____
b) 71.524 is closer to ____ than to ____
c) 71.524 rounds to ____ (to 2 decimal places).

2) a) Sketch a number line from 3.45 to 3.46
b) Mark **3.458** on your number line.
c) Does 3.458 round to 3.45 or 3.46 (to 2 decimal places)?

3) a) Copy the number 9.461.
Circle the third digit after the decimal point.
b) Does 9.461 round down or round up?
c) Round 9.461 to 2 decimal places.

4) Use the method in question 3 to round these numbers to 2 decimal places:
a) 23.714
b) 18.526
c) 134.629
d) 81.365
e) 7.009
f) 81.365

5) A concrete block weighs 17.948 kg.
Round its mass to:
a) 2 decimal places
b) 1 decimal place
c) the nearest kilogram.

Decimals to 1, 2 and 3 decimal places

Adding and subtracting thousandths and numbers with different decimal places

 I will learn to add and subtract decimal fractions with 1, 2 and 3 decimal places.

A number with **three digits after the decimal point** (tenths, hundredths and thousandths) is written to **3 decimal places**.

Sometimes you may have to add or subtract decimals with thousandths. For example:

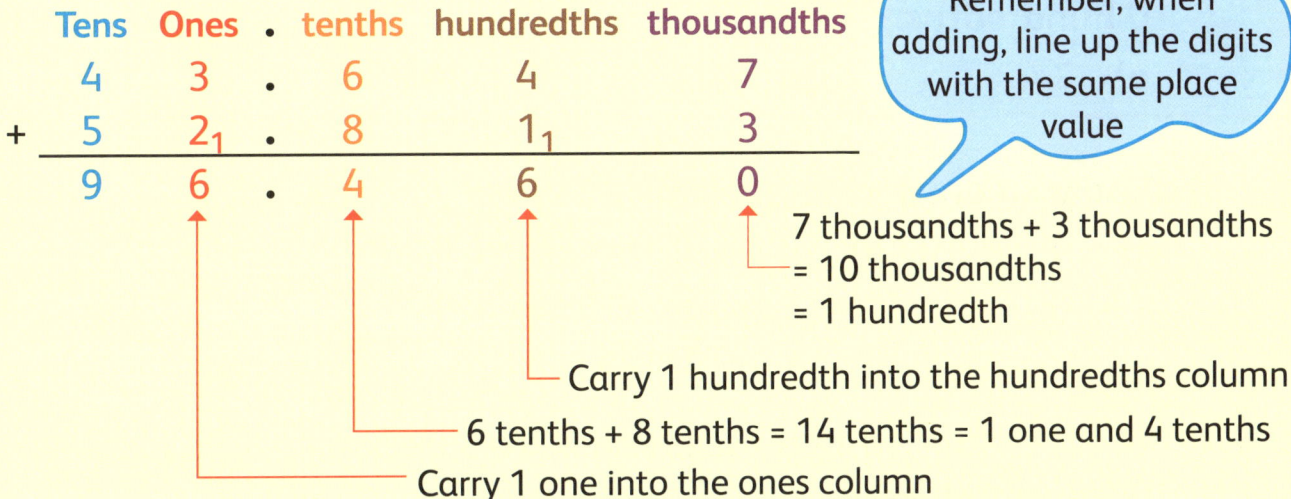

Remember, when adding, line up the digits with the same place value

7 thousandths + 3 thousandths = 10 thousandths = 1 hundredth

Carry 1 hundredth into the hundredths column

6 tenths + 8 tenths = 14 tenths = 1 one and 4 tenths

Carry 1 one into the ones column

Sometimes you may have to add or subtract decimals with different numbers of decimal places.
For example:
9.8 − 4.62

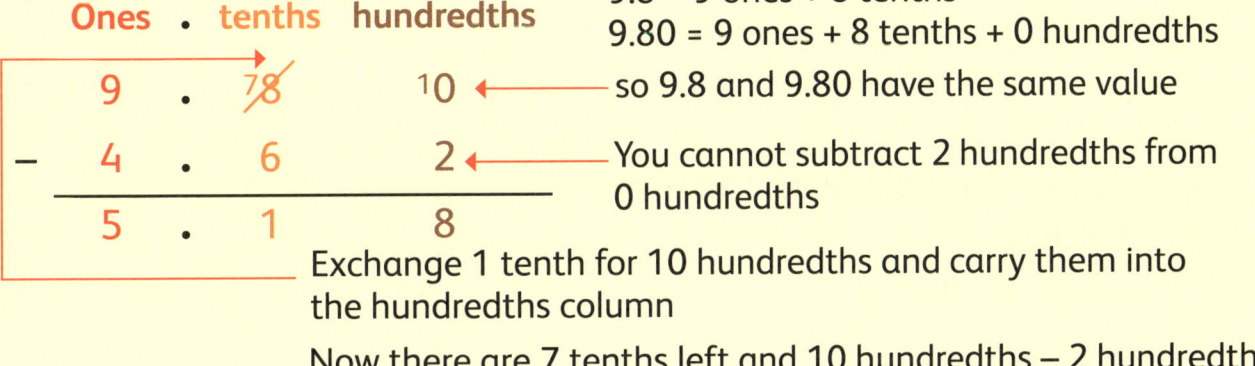

9.8 = 9 ones + 8 tenths
9.80 = 9 ones + 8 tenths + 0 hundredths
so 9.8 and 9.80 have the same value

You cannot subtract 2 hundredths from 0 hundredths

Exchange 1 tenth for 10 hundredths and carry them into the hundredths column

Now there are 7 tenths left and 10 hundredths − 2 hundredths

Chapter 7 Decimal fractions 1

Exercise 4

1) Work out:

   a)   621.047
         + 168.912
       _____

   b)   3.615
         − 2.103
       _____

   c)   84.362
         + 19.157
       _____

   d)   92.884
         − 37.126
       _____

2) In your jotter, write these numbers to 2 decimal places with **0 hundredths**:

   a) 7.4   b) 41.3   c) 39.7   d) 415.9

3) Use your answers to question 2 to help you work out:

   a)   7.4
      + 6.24
      _____

   b)   2.71
      + 41.3
      _____

   c)   39.7
      − 19.15
      _____

   d)   415.9
      − 98.43
      _____

4) In your jotter, write these numbers to 3 decimal places with **0 thousandths**:

   a) 4.82   b) 9.17   c) 16.50   d) 23.01

5) Use your answers to question 4 to help you work out:

   a)   4.82
      + 3.265
      _____

   b)   14.356
      − 9.17
      _____

   c)   16.5
      + 2.629
      _____

   d)   23.01
      − 7.309
      _____

6) Line up the digits with the same place value to work out these additions:

   a) 15.3 + 24.19   b) 1.267 + 5.5
   c) 1.98 + 18.6   d) 392.78 + 21.506

Decimals to 1, 2 and 3 decimal places

7) Line up the digits with the same place value to work out these subtractions:
   a) 17.83 – 1.2
   b) 9.627 – 1.54
   c) 28.4 – 5.79
   d) 1047.62 – 238.102

8) A lorry is 9.5 m long.
A car is 4.36 m long.
How much **longer** is the lorry than the car?

9) Sam cycles 8.65 km in the morning.
He cycles 10.5 km in the afternoon.
How many kilometres does he cycle **altogether**?

10) The length of a classroom is 12.95 m.
The width of a classroom is 7.6 m.
What is the perimeter of the classroom?

Now try this!

Work with a partner.
Mo says:

> I am thinking of a number.
> It has 3 decimal places.
> It lies between 8.43 and 8.44
> When rounded to 2 decimal places it rounds up to 8.44

Guess Mo's number.
Is this the only possibility for Mo's number?

Chapter 7 Decimal fractions 1

Revisit, review, revise

1) Which number is written to 3 decimal places?

| 0.3 | 30.5 | 6.194 | 4.53 |

2) In the decimal **9035.642**, what does the:
   a) 9 represent
   b) 2 represent
   c) 3 represent
   d) 4 represent
   e) 5 represent
   f) 6 represent?

3) Mentally work out what number is:
   a) 0.002 less than 29.075
   b) 0.002 more than 29.075

4) In your jotter, write the **biggest** number in each group.
   a) 10.5, 10.9, 11.1
   b) 481.36, 481, 481.5

5) In your jotter, write each group of numbers in order of size. Start with the **smallest**.
   a) 9.31, 9.7, 8.4
   b) 0.602, 6.2, 0.062

6) a) Imagine a number line from 15.7 to 15.8
   b) Imagine **15.74** on your number line.
   c) Does 15.74 round to 15.7 or 15.8 (to 1 decimal place)?

7) Round these numbers to 1 decimal place:
   a) 9.27
   b) 18.88
   c) 5.02
   d) 409.75

8) Round these numbers to 2 decimal places:
   a) 4.192
   b) 73.157
   c) 2.804
   d) 0.065

9) A tree is 17.648 m tall.
   Round its height to 2 decimal places

10) Work out:
   a) 63.047 + 19.925
   b) 0.842 − 0.671
   c) 52.6 + 31.73
   d) 407.8 − 86.01

8 Angles
Working with angles

Measuring and classifying angles

 I will learn to classify and measure angles.

Remember, remember

An angle is a measure of turn.
Angles are measured in degrees.

A full turn is 360°.

A half turn, or the **angle on a straight line** is 180°.

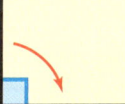

A quarter turn, or **right angle**, is 90°.
A right angle is indicated with a small square.

An **acute** angle is between 0° and 90°.

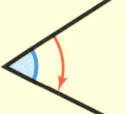

An **obtuse** angle is between 90° and 180°.

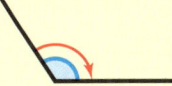

A **reflex** angle is between 180° and 360°.

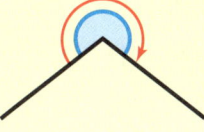

Chapter 8 Angles

Use a **protractor** to measure an angle.

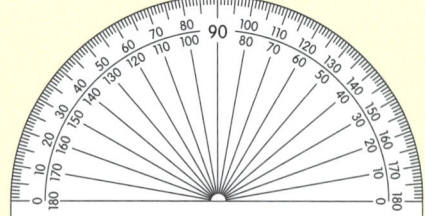

Step 1: place the centre of the protractor on the centre of the angle you are measuring.

Step 2: turn the protractor until the zero line lies along one arm of the angle.

Step 3: count from the zero (inside or outside) and read the value where the other arm cuts the scale.

Example

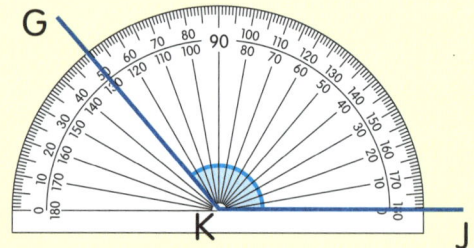

Use inside scale <GKJ = 130°

Always **estimate** the size of the angle before measuring. This helps you to check that your answers are sensible.

Remember, remember

You name an angle using three capital letters.
The vertex is the middle letter.
The angle in the box above is called <GKJ or <JKG.

Working with angles

Exercise 1

1) | 180° 47° 4120° 495° 490° 468° 199° 359° |

Which of the above angles are:
a) acute
b) obtuse
c) right angles
d) straight
e) reflex?

2) Measure each of the angles.

a) <ACB

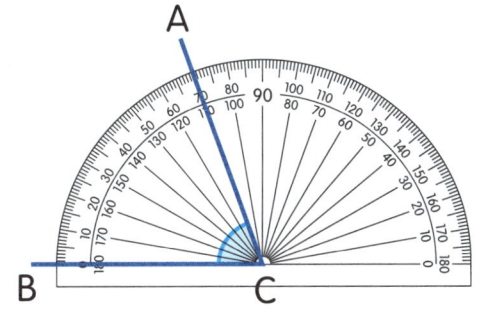

b) <DFE

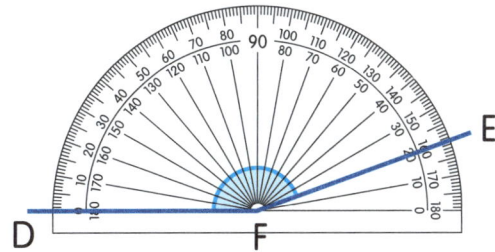

c) <IHG

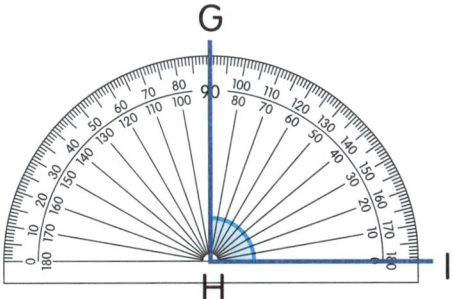

d) <JLK

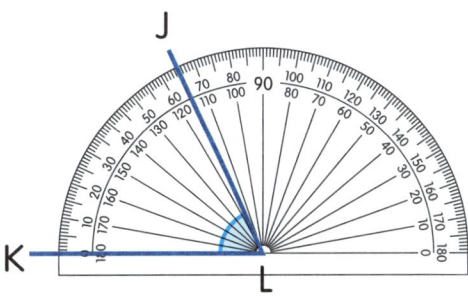

e) <NOM

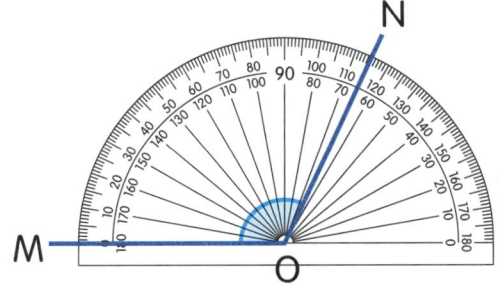

f) <UST

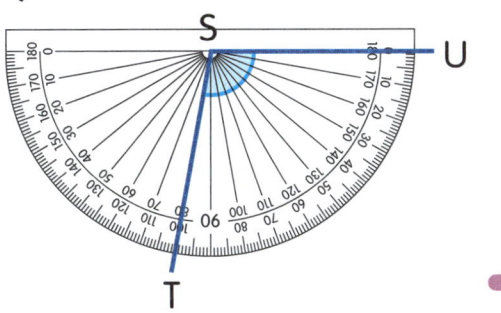

Chapter 8 Angles

g) <VWX

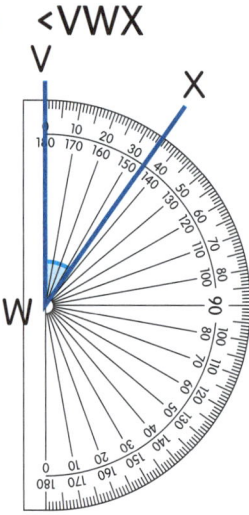

h) <YAZ

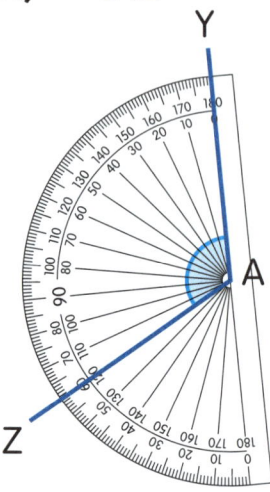

3) Which angles in question 2 are:

a) acute

b) obtuse?

4) Do not use a protractor in this question.
Choose the estimate closest to what you think the angle is.

a) 40°, 60° or 88°

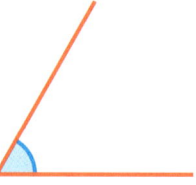

b) 10°, 30° or 70°

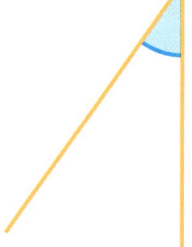

c) 60°, 85° or 110°

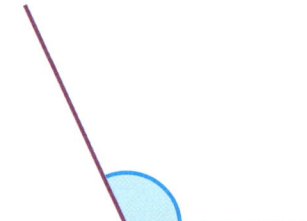

d) 100°, 140° or 170°

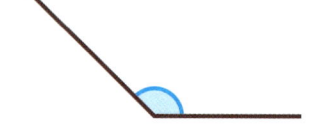

e) 15°, 50° or 75°

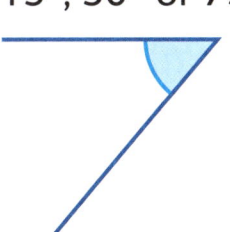

f) 100°, 120° or 160°

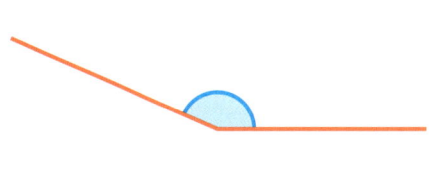

5) Use a protractor to measure each angle and write them in your jotter.

a)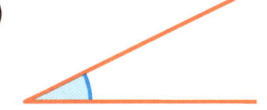

b)

c)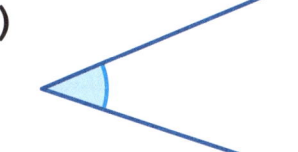

d)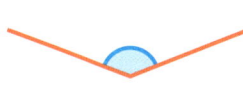

e)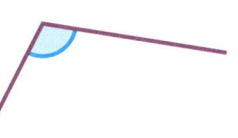

f)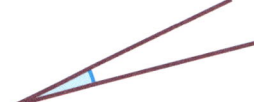

g)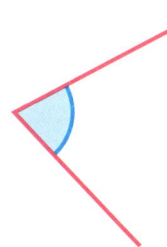

h)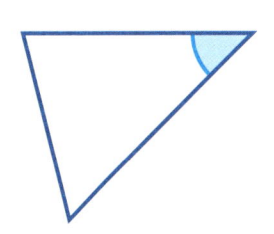

6) Abdul thinks that this angle measures 40°.
Is Abdul correct?
Explain your answer.

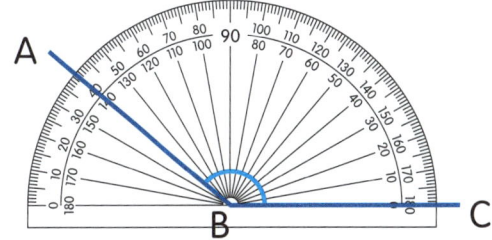

Chapter 8 Angles

Constructing angles

 I will learn to construct angles using a ruler and protractor.

To construct an angle, you will need to use a ruler, protractor and pencil.

Example

Construct <TPQ = 40°.

Step 1: Draw a line with a dot at one end.

Step 2: Put the centre of the protractor on the dot and line up with the line.

Step 3: Count round from the zero line up to the 40° mark and mark with a dot.

Step 4: Join the dots and write in the letters (middle letter P).

Step 1

Step 2

Step 3

Step 4

Working with angles

Exercise 2

 1) a) Copy this line.

B ———————— C

b) Use a protractor to construct <ABC = 30°.

2) Use a protractor to construct and label these angles:
- a) <ABC = 50°
- b) <PQR = 90°
- c) <JKL = 10°
- d) <DEF = 110°
- e) <XYZ = 175°
- f) <PQR = 45°

 3) Use a protractor to construct and label these angles:
- a) <ABC = 38°
- b) <XYZ = 22°
- c) <PQR = 96°
- d) <DEF = 108°
- e) <JKL = 123°
- f) <VWX = 8°

4) Amir says that he can't draw an angle larger than 180° because his protractor only goes up to that angle.

Is Amir correct?

Explain.

Now try this!

Work with a partner.

Without using a protractor, you should both draw an angle as close to 80° as you can.

Swap with your partner and measure their angle.

The person whose angle is closest to 80° wins a point.

Choose a different size angle and repeat.

The first person to 5 points wins.

Chapter 8 Angles

Revisit, review, revise

1) Estimate the size of each angle, then measure it with a protractor.

a)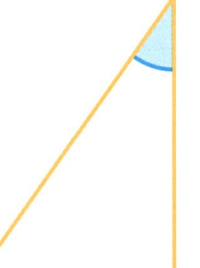

b)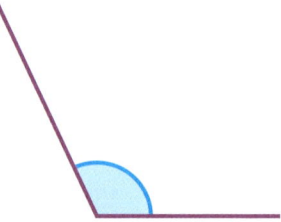

c)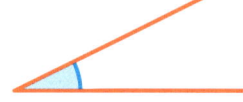

d)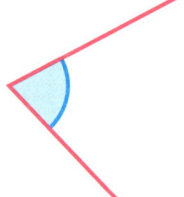

e)

f)

g)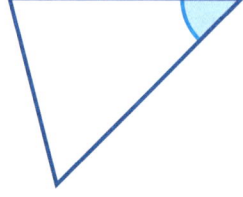

h)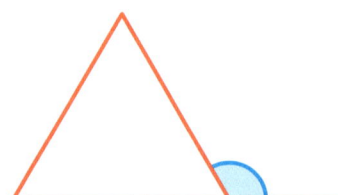

2) Use a protractor to construct and label these angles:

a) <ABC = 20°
b) <DEF = 60°
c) <XYZ = 110°
d) <JKL = 170°
e) <PQR = 75°
f) <VWX = 142°

9 Decimal fractions 2
Multiplying and dividing decimals by 10, 100 and 1000

Multiplying decimals by 10, 100 and 1000

 I will learn how to multiply a decimal by 10, 100 or 1000.

Remember, remember

When you **multiply by 10**:
- digits move **1 place left** and the space is filled with a zero
- the number becomes **10 times bigger**.

For example:

	Hundreds	Tens	Ones
32 × 10 =		3	2
	3	2	0

When you **multiply by 100**:
- digits move **2 places left** and spaces are filled with a zero
- the number becomes **100 times bigger**.

For example:

	Thousands	Hundreds	Tens	Ones
51 × 100 =			5	1
	5	1	0	0

When you **multiply by 1000**:
- digits move **3 places left** and spaces are filled with a zero
- the number becomes **1000 times bigger**.

For example:

	Thousands	Hundreds	Tens	Ones
6 × 1000 =				6
	6	0	0	0

Chapter 9 Decimal fractions 2

When you **multiply a decimal by 10, 100 or 1000**, the same happens as when you multiply whole numbers by 10, 100 or 1000.

When you **multiply a decimal by 10**:
- digits move **1 place left**
- the number becomes **10 times bigger**.

For example:

4.37 × 10 =

Tens	Ones	. tenths	hundredths
	4	3	7
4	3	. 7	

When you **multiply a decimal by 100**:
- digits move **2 places left**
- the number becomes **100 times bigger**.

For example:

21.32 × 100 =

Thousands	Hundreds	Tens	Ones	. tenths	hundredths
		2	1	. 3	2
2	1	3	2	.	

72.8 × 100 =

Thousands	Hundreds	Tens	Ones	. tenths
		7	2	. 8
7	2	8	0	.

In the second example, the movement leaves a space in the ones column. Spaces between digits and the decimal point are filled with a zero.

When you **multiply a decimal by 1000**:
- digits move **3 places left**
- the number becomes **1000 times bigger**.

Multiplying and dividing decimals by 10, 100 and 1000

For example:

0.84 × 1000 =

Thousands	Hundreds	Tens	Ones	.	tenths	hundredths
			0	.	8	4
	8	4	0	.		

In this example, the movement leaves a space in the ones column, which is filled with a zero.

Also, the movement would move the 0 of 0.84 into the thousands column. Zeroes at the start of numbers are not necessary, unless they are in the ones column.

Exercise 1

 1) Copy and complete these place value tables:

a)

52.61 × 10 =

Thousands	Hundreds	Tens	Ones	.	tenths	hundredths	thousandths
		5	2	.	6	1	
	5			.			

b)

7.59 × 100 =

Thousands	Hundreds	Tens	Ones	.	tenths	hundredths	thousandths
			7	.	5	9	
	7			.			

c)

4.82 × 1000 =

Thousands	Hundreds	Tens	Ones	.	tenths	hundredths	thousandths
			4	.	8	2	
4				.			

2) Work out:
- a) 6.1 × 10
- b) 9.25 × 10
- c) 74.1 × 10
- d) 86.32 × 10
- e) 974.6 × 10
- f) 285.04 × 10
- g) 3.65 × 100
- h) 48.91 × 100
- i) 1.972 × 1000

Chapter 9 Decimal fractions 2

3) Work out:
a) 0.32 × 10
b) 0.94 × 100
c) 1.7 × 100
d) 16.8 × 100
e) 0.28 × 1000
f) 6.3 × 1000

4) What numbers are **missing**?
a) 4.31 × ____ = 431
b) 7.85 × ____ = 78.5
c) ____ × 100 = 605
d) ____ × 10 = 97.84
e) ____ × 1000 = 553
f) 3.09 × ____ = 3090

Remember, remember

1 cm = 10 mm
To convert from **centimetres** to **millimetres** multiply by 10.

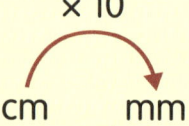

1 m = 100 cm
To convert from **metres** to **centimetres** multiply by 100.

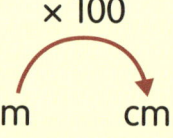

1 km = 1000 m
To convert from **kilometres** to **metres** multiply by 1000.

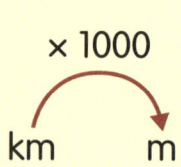

5) Copy and complete these measures:
a) 5.7 cm is ____ mm
b) 0.73 cm is ____ mm
c) 87.29 m is ____ cm
d) 46.1 m is ____ cm
e) 9.28 km is ____ m
f) 5.6 km is ____ m.

Multiplying and dividing decimals by 10, 100 and 1000

Dividing decimals by 10, 100 and 1000

 I will learn how to divide a decimal by 10, 100 or 1000. I will also learn how to divide a whole number by 10, 100 or 1000 with a decimal answer.

Remember, remember

When you **divide by 10**:
- digits move **1 place right**
- the number becomes **10 times smaller**.

For example:

680 ÷ 10 =

Hundreds	Tens	Ones
6	8	0
	6	8

When you **divide by 100**:
- digits move **2 places right**
- the number becomes **100 times smaller**.

For example:

9300 ÷ 100 =

Thousands	Hundreds	Tens	Ones
9	3	0	0
		9	3

When you **divide by 1000**:
- digits move **3 places right**
- the number becomes **1000 times smaller**.

For example:

8000 ÷ 1000 =

Thousands	Hundreds	Tens	Ones
8	0	0	0
			8

When you **divide a decimal by 10, 100 or 1000**, the same happens as when you divide whole numbers by 10, 100 or 1000.

When you **divide a decimal by 10**:

- digits move **1 place right**
- the number becomes **10 times smaller.**

For example:

25.7 ÷ 10 =

Tens	Ones	• tenths	hundredths
2	5	• 7	
	2	• 5	7

When you **divide a decimal by 100**:

- digits move **2 places right**
- the number becomes **100 times smaller.**

For example:

1673.4 ÷ 100 =

Thousands	Hundreds	Tens	Ones •	tenths	hundredths	thousandths
1	6	7	3 •	4		
		1	6 •	7	3	4

37.6 ÷ 100 =

Tens	Ones	• Tenths	hundredths	thousandths
3	7	• 6		
	0	• 3	7	6

In the second example, the movement leaves a space in the ones column.

Zeroes at the start of numbers are not necessary, unless they are in the ones column.

When you **divide a decimal by 1000**:

- digits move **3 places right**
- the number becomes **1000 times smaller.**

Multiplying and dividing decimals by 10, 100 and 1000

For example:
8620 ÷ 1000 =

Thousands	Hundreds	Tens	Ones ·	tenths	hundredths	thousandths
8	6	2	0 ·			
			8 ·	6	2	

In this example, the movement leaves a zero in the thousandths column. Zeroes at the end of decimals are not usually necessary.

Exercise 2

1) Copy and complete these place value tables:

   a)
   64.81 ÷ 10 =

Thousands	Hundreds	Tens	Ones ·	tenths	hundredths	thousandths
		6	4 ·	8	1	
			6 ·			

   b)
   3716.8 ÷ 100 =

Thousands	Hundreds	Tens	Ones ·	tenths	hundredths	thousandths
3	7	1	6 ·	8		
			·	1		

   c)
   924 ÷ 1000 =

Thousands	Hundreds	Tens	Ones ·	tenths	hundredths	thousandths
	9	2	4 ·			
			0 ·			

2) Work out:
   a) 38.7 ÷ 10
   b) 549.16 ÷ 10
   c) 1967 ÷ 10
   d) 1863.3 ÷ 100
   e) 209.4 ÷ 100
   f) 9746 ÷ 1000

3) Work out:
   a) 4.6 ÷ 10
   b) 178 ÷ 1000
   c) 2970 ÷ 1000

Chapter 9 Decimal fractions 2

4) Copy and complete:
   a) 1 ÷ 10 = 0.____   b) 3 ÷ 100 = 0.0____   c) 7 ÷ 1000 = 0.00____

5) Work out:
   a) 5.9 ÷ 100   b) 26 ÷ 1000   c) 9.04 ÷ 10

6) What numbers are **missing**?
   a) 984 ÷ ____ = 9.84   b) 29.52 ÷ ____ = 2.952
   c) ____ ÷ 100 = 6.053   d) ____ ÷ 1000 = 0.482

Remember, remember

1 cm = 10 mm
To convert from **millimetres** to **centimetres** divide by 10.

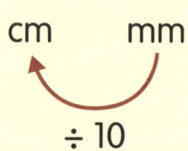

1 m = 100 cm
To convert from **centimetres** to **metres** divide by 100.

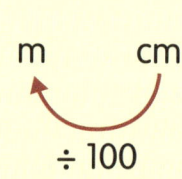

1 km = 1000 m
To convert from **metres** to **kilometres** divide by 1000.

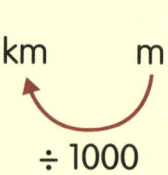

7) Copy and complete these measures:
   a) 35 mm is ____ cm   b) 738 mm is ____ cm
   c) 187.2 cm is ____ m   d) 46.1 cm is ____ m
   e) 2285 m is ____ km   f) 741 m is ____ km

8) Sara says 9000 cm is 0.9 km.
   Is she correct?
   Show your working.

Multiplying and dividing decimals by 10, 100 and 1000

Now try this!

Work with a partner.

Find all the different ways to copy and complete these calculations with:

| 0.06 | 0.6 | 6 | or | 60 |

1) ____ × 10 = ____   2) ____ × 100 = ____   3) ____ × 1000 = ____
4) ____ ÷ 10 = ____   5) ____ ÷ 100 = ____   6) ____ ÷ 1000 = ____

Revisit, review, revise

1) Work out:
   a) 463.8 × 10   b) 5.17 × 100   c) 0.37 × 10
   d) 4.9 × 100   e) 1.5 × 1000   f) 0.67 × 1000

2) Work out:
   a) 42.8 ÷ 10   b) 5073 ÷ 1000   c) 129.7 ÷ 100
   d) 37.1 ÷ 100   e) 592 ÷ 1000   f) 49 ÷ 1000

3) What numbers are **missing**?
   a) 5.19 × ____ = 519   b) 7.84 × ____ = 78.4
   c) ____ × 100 = 83.7   d) 16.34 ÷ ____ = 1.634
   e) ____ ÷ 100 = 4.78   f) ____ ÷ 1000 = 0.65

4) Copy and complete:
   a) £393.12 is ____ p   b) 1035p is £ ____   c) 6.1 cm is ____ mm
   d) 93.05 m is ____ cm   e) 39.4 m is ____ cm f) 4.8 km is ____ m
   g) 85 mm is ____ cm   h) 945 m is ____ km.

5) 100 people receive an equal share of £9475.
   How much do they each receive?
   In your jotter, write your answer in £.

10 Money
Profit, loss and budgeting

Profit and loss

> I will learn to calculate profit and loss.

When you sell something for more than it costs you to buy it or make it, there is a **profit**.

profit = selling price – buying (or making) price

For example:

Sam buys a mobile phone for £150.

Later, he sells it for £240.

Sam sells his mobile phone for **more** than it cost to buy it:

Profit = £240 – £150 = £90

Sam makes a **profit** of £90.

When you sell something for less than it costs you to buy it or make it, there is a **loss**.

loss = buying (or making) price – selling price

For example:

Ola sells a cake for £4.

The cake ingredients cost £4.50.

Ola sells the cake for **less** than the cost of the ingredients:

Loss = £4.50 – £4 = £0.50

Ola makes a **loss** of £0.50.

Profit, loss and budgeting

Exercise 1

1) Alice buys and sells secondhand clothes.
   In your jotter, write whether Alice makes a profit or loss on each item.
   a) Jeans: buys for £7, sells for £12
   b) Coat: buys for £36, sells for £28
   c) T-shirt: buys for £2.80, sells for £3
   d) Belt: buys for £6.20, sells for £4.80

2) For each item in question 1, work out how much profit or loss Alice makes.
   In your jotter, write: profit = _____ or loss = _____

3) On which item of clothing in question 1 does Alice make:
   a) the biggest profit
   b) the biggest loss?

4) Rashana buys a signed football shirt for £250.
   She sells it a year later for £345.
   a) Does she make a profit or loss?
   b) How much profit or loss does she make?

5) Cameron buys a car for £8750.
   He sells it for £5475.
   How much loss does he make?

6) Sari buys two paintings for a **total** of £1050.
   She sells one for £830 and the other for £650.
   How much profit does Sari make **altogether**?

Chapter 10 Money

7) A shopkeeper buys a box of 10 glasses for £32.50.

He sells **each** glass for £4.50.

After selling all 10 glasses:

a) does he make a profit or loss

b) how much profit or loss does he make?

 8) Rani makes a cupboard.

Here are her costs:

£370 for wood   £2.60 for screws   £24.35 for paint

Rani wants to make a profit of at least £100.

What is the **minimum** (smallest) amount Rani can sell the cupboard for?

Budgeting

💡 I will learn to solve problems involving budgeting.

When you plan what money to keep and what money to spend, it is called **budgeting**.

If you have enough money to pay for something, then you **can afford** it.

If you do not have enough money to pay for something, then you **cannot afford** it.

Example

Ivan is paid £550 each week.

Each week, he budgets:

£200 for rent   £160 for food   £70 for bus fare   £50 savings

a) What is Ivan's total budget each week?
b) How much does Ivan have left each week?
c) Can Ivan afford to buy a concert ticket for £100 this week? Yes or no.

Answers

a) Ivan budgets: £200 + £160 + £70 + £50

You can work this out mentally, counting up in hundreds and tens, or you can use column addition:

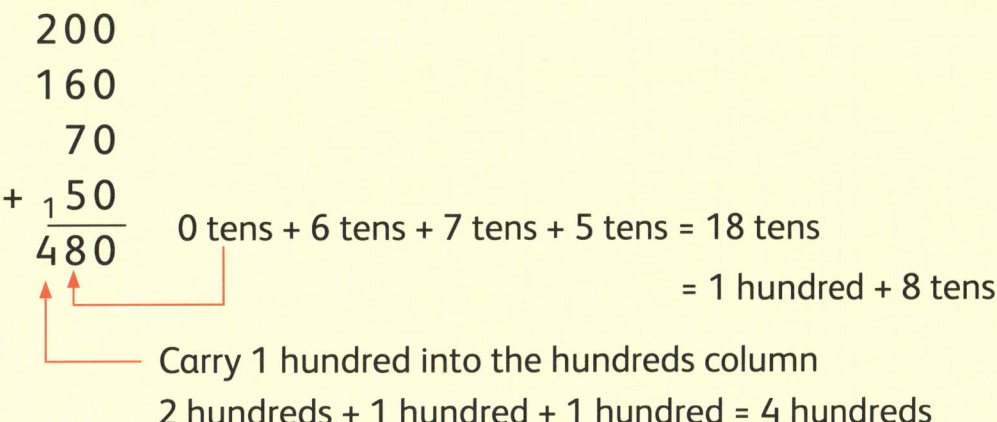

0 tens + 6 tens + 7 tens + 5 tens = 18 tens

= 1 hundred + 8 tens

Carry 1 hundred into the hundreds column

2 hundreds + 1 hundred + 1 hundred = 4 hundreds

Ivan budgets £480.

b) Ivan has left: £550 − £480

You can work this out mentally, counting back from £550 to £480, or you can use column subtraction:

You cannot subtract 8 tens from 5 tens

Exchange 1 hundred for 10 tens and carry them into the tens column

Now there are 4 hundreds left and 15 tens − 8 tens

Ivan has £70 left.

c) No, Ivan cannot afford the concert ticket this week. Ivan has £70 left but the concert ticket is £100.

Chapter 10 Money

Exercise 2

 1) Here is Billy's holiday budget:

hotel: £500 food: £250 spending money: £450

What is Billy's **total** budget for his holiday?

 2) Bella is saving for a car.
She has £3600.
She sees a car for sale for £3025.
Can Bella afford the car? Yes or no.

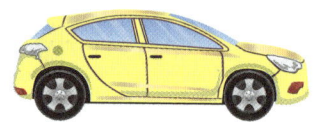

3) Ali is saving for a record player and speakers.
He has £510.
The record player is £365 and the speakers are £155.
Can Ali afford to buy the record player and speakers? Yes or no.
Show your working.

4) Josh is saving for hiking boots.
They cost £150.
He saves £12 every week.
  a) Can he afford the hiking boots after 9 weeks? Yes or no.
  b) How many **more** weeks does he need to save?

5) Polly has a new puppy.
For the first month, she budgets:

vets £60 insurance £23 food £40

Polly spends £134 in the first month.
  a) Does she spend more or less than her budget?
  b) How much more or less?

Profit, loss and budgeting

 6) A dance class costs £6.

The teacher budgets £28 for the hire of the dance hall.

How many people need to come to the class for the teacher to make a profit?

Now try this!

Javi works 5 days a week.

Each week he spends money on train fares and lunches.

He writes a budget like this:

> Budget per week: £60
> £40 for train fares to and from work.
> £5 a day for lunch.

Javi's budget doesn't work.

Re-write it a different way so that it works.

Revisit, review, revise

1) Cara makes pottery.
   a) She makes vases for £5.25 and sells them for £13.75
   b) She makes plates for £3.05 and sells them for £4.15
   c) She makes mugs for £3.50 and sells them for £2.70

   In your jotter, write each item of pottery and:
   i) whether Cara makes a profit or loss
   ii) how much profit or loss she makes.

2) On which item in question 1 does Cara make the **biggest profit**?

Chapter 10 Money

3) Mirabelle buys a bicycle for £247.

   She sells it for £180.

   a) Does she make a profit or loss?

   b) How much profit or loss does she make?

4) Carter buys an autograph by a famous footballer for £412.

   He sells it online for £825.

   How much profit does he make?

5) Caitlin wants to go to London for her birthday.

   Her parents give her £150, her Gran gives her £55 and her brother gives her £20.

   The trip to London costs £220.

   Can she afford to go? Yes or no.

   Show your working.

6) Antony is saving for a trumpet.

   The trumpet costs £898.

   He saves £120 each month.

   a) Can he afford the trumpet after 6 months? Yes or no.

   Show your working.

   b) How many more months does he need to save?

7) Farah makes children's toys.

   For one toy she budgets:

   £5.60 for material   £0.80 for paint   £0.25 for glue

   Farah sells each toy for £12.50.

   How much profit does she make on each toy?

2D shapes
Triangles, squares and rectangles

Types of triangle

💡 I will learn the properties of triangles.

Triangles are three sided polygons. There are 4 types of triangle:

Right-angled triangles:
one angle is a right angle (90° angle).

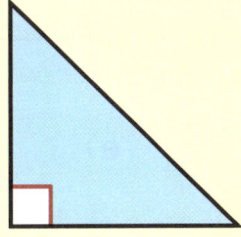

Isosceles triangles: two angles are equal and two sides are the same length.

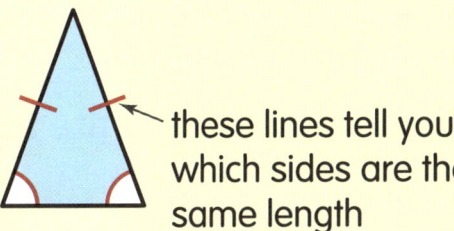

← these lines tell you which sides are the same length

Equilateral triangles: all sides are equal and all angles are equal.

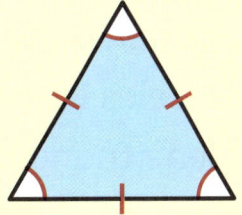

Scalene triangles: no sides are equal and no angles are equal.

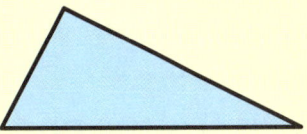

Chapter 11 2D shapes

Exercise 1

1) What types of triangles are these?
   Choose from: right-angled triangle, isosceles triangle, equilateral triangle, scalene triangle.

   a) 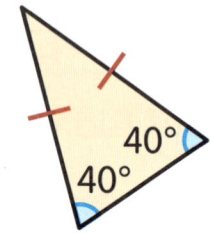   b) 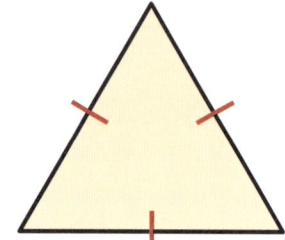   c)

   d) 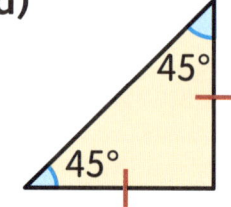   e) 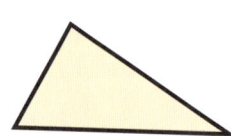   f)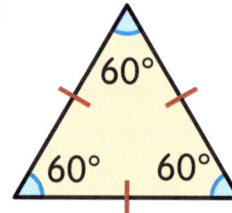

2) In your jotter, write the name of the triangle being described.
   a) All sides are equal.
   b) It contains a 90° angle.
   c) It has no equal sides.
   d) Two angles are equal.

3) On dotty paper, draw and label:
   a) an isosceles triangle
   b) a right-angled triangle
   c) a scalene triangle
   d) an equilateral triangle.

Polygons are often named using the letters on their vertices.
This is triangle JKL.
Triangle JKL is scalene **and** right-angled.

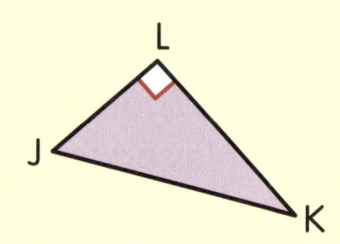

Triangles, squares and rectangles

4) Harim says that triangle ABC is an isosceles triangle.
India says that it is a right-angled triangle.
Who is correct?
Explain your answer.

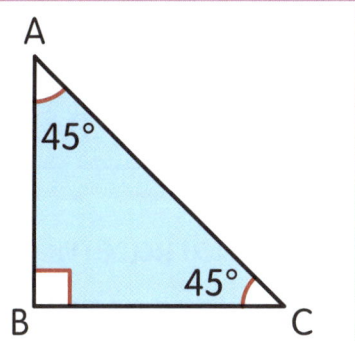

5) Copy or trace these triangles.
Draw on any lines of symmetry.

a)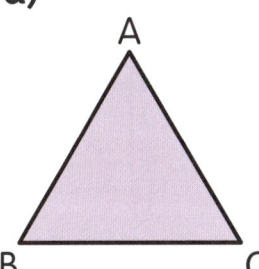
b)
c)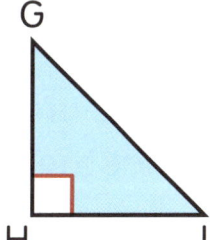
d)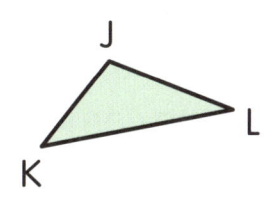

6) Draw a triangle ABC with no lines of symmetry.
What type of triangle is it?

7) a) Draw a triangle PQR that is both right-angled and scalene.
   b) Draw a triangle XYZ that is both right-angled and isosceles.

8) Can a triangle contain more than one right angle?
Try drawing it.

Chapter 11 2D shapes

Squares and rectangles

 I will learn the properties of squares and rectangles.

A **quadrilateral** is a 4-sided shape.
The sides are all straight lines.
The sides join (there are no gaps).
It is called a **closed shape**.

A **square** is a quadrilateral. ABCD is a square.
All the sides are equal in length: AB = BC = CD = DA.
The lines on the sides show that they are all the same length.
All the angles in a square are 90°. The squares in the corners show that the angles are all 90°.

A **rectangle** is a quadrilateral. EFGH is a rectangle.
The opposite sides are equal in length.
- EH = FG. The single lines show that these sides are equal.
- EF = HG. The double lines show that these sides are equal.

All the angles in a rectangle are 90°.

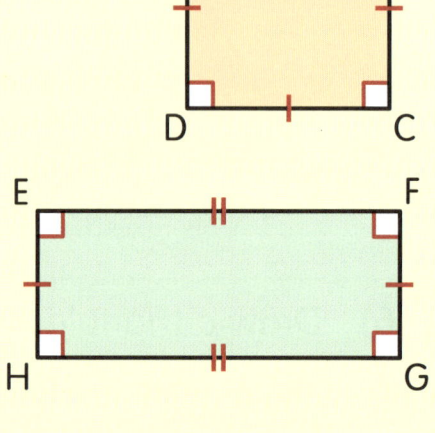

Two lines are **parallel** if they always have the same distance between them.
The lines XY and MN are parallel.
Parallel lines **do not** have to be the same length.
Railway tracks are always parallel; the two tracks must be the width of the train apart.

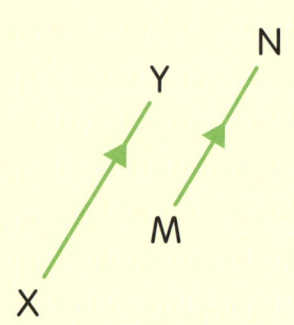

Triangles, squares and rectangles

Exercise 2

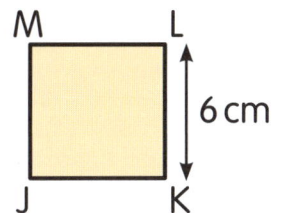

1) JKLM is a square.
   a) How long is side:
      i) ML ii) JK iii) JM?
   b) What is the size of:
      i) <MLK ii) <LKJ
      iii) <KJM iv) <JML?

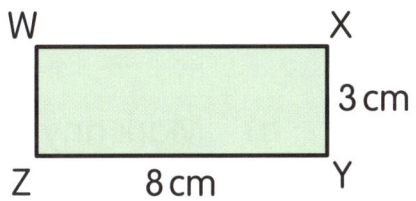

2) WXYZ is a rectangle.
   a) How long is side:
      i) ZW
      ii) WX?
   b) What is the size of:
      i) <WXY ii) <XYZ
      iii) <YZW iv) <ZWX?

3) a) On squared paper draw a square ABCD with sides 5 cm long.
   b) Mark any sides which are the same length using l
   c) Mark any right angles using □
   d) Which side is parallel to AB?
   e) Which side is parallel to BC?

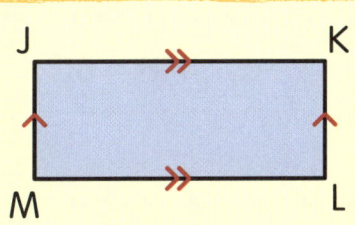

You indicate parallel sides on a shape with arrows.
LK is parallel to MJ.
JK is parallel to ML.

Chapter 11 2D shapes

4) On the square you drew for question 3, mark the parallel sides.

5) a) On squared paper, copy rectangle DEFG, with sides 9 cm and 4 cm.

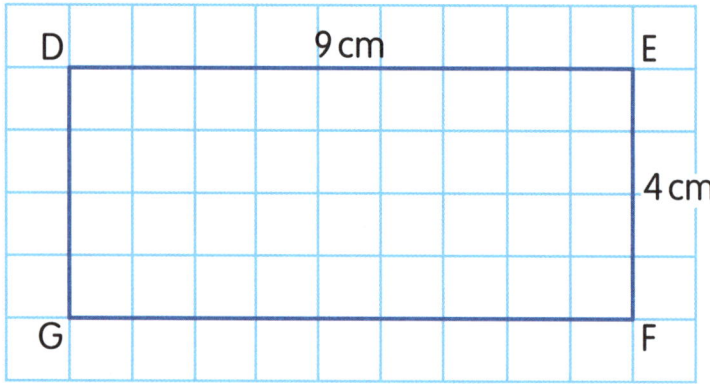

b) Mark any sides the same length using I and II
c) Mark any right angles using □
d) Mark any parallel sides using ⌃ or ⌃⌃

 6) How many lines of symmetry do these shapes have?
a) a square
b) a rectangle

 7) ABCD is a rectangle.

Are these statements true or false?
a) All the sides are the same length.
b) <ABC = <ADC
c) AB is parallel to DC.
d) AB = BC

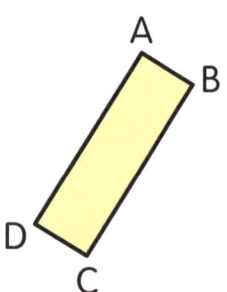

Triangles, squares and rectangles

Now try this!

Work with a partner.

On squared paper draw a square and a rectangle.

Draw diagonal lines joining the opposite corners.

Measure the angles between the lines using a protractor.

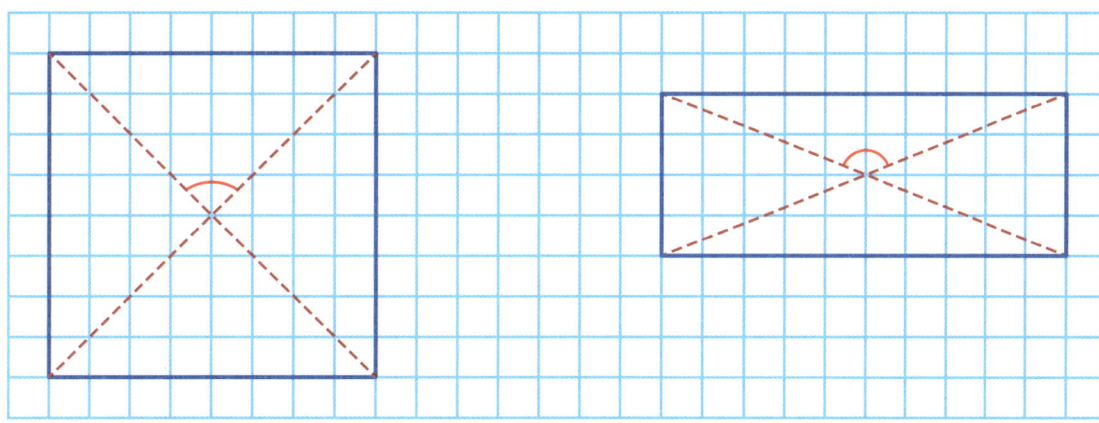

Try for different squares and rectangles. What do you notice?

Revisit, review, revise

1) What type of triangles are these?

   Choose from right-angled triangle, isosceles triangle, equilateral triangle, scalene triangle.

   a) 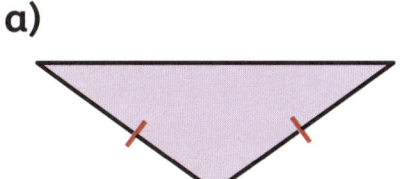   b) 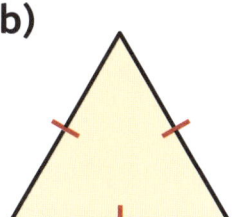   c) 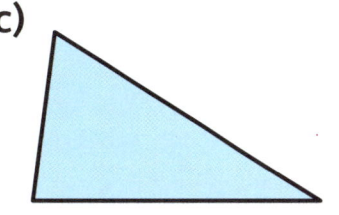   d)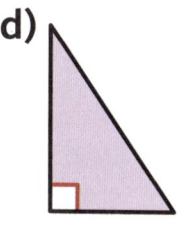

2) Can a triangle be:
   a) isosceles and right-angled
   b) equilateral and right-angled
   c) right-angled and scalene
   d) isosceles and scalene?

   Draw them to check your answers.

3) ABCD is a square. EFGH is a rectangle.

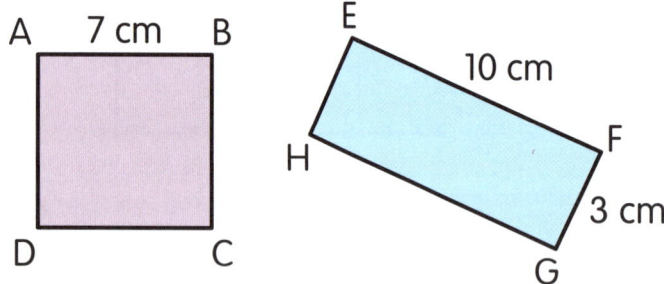

   a) What is the length of:
      i) BC   ii) EH   iii) AD   iv) GH?
   b) What is the size of:
      i) <ABC   ii) <HEF   iii) <GFE   iv) <ADC?
   c) Which side is parallel to:
      i) AB   ii) BC   iii) GH   iv) FG?

12 Equations
Introduction to algebra

Greater than, less than, equal to, not equal to

 I will learn to recognise and understand the symbols <, >, = and ≠

Remember, remember

You can compare numbers using **symbols**:

- < means **less than**
- > means **greater than**.

The pointed part of the symbol < or > always points to the smaller number.

7 is greater than 3   4 is less than 9
7 > 3   4 < 9

= means **is equal to** (has the same value).
≠ means **is not equal to**.
1 + 7 is not equal to 9
1 + 7 ≠ 9

Example

Which symbol < or > should be put between these calculations?
3 × 9   4 × 6

Answer
3 × 9 > 4 × 6
(27   >   24)

Chapter 12 Equations

Exercise 1

 1) Here are 10 calculations.
Match the pairs with an = sign between them.

| 8 × 3 | 22 – 10 | 21 ÷ 3 | 18 ÷ 2 | 27 ÷ 3 |
| 18 – 11 | 15 + 9 | 26 + 24 | 6 × 2 | 25 × 2 |

2) Copy the calculations and put the symbol = or ≠ between them.
a) 6 + 4 ___ 2 + 8
b) 10 – 6 ___ 14 – 11
c) 5 × 6 ___ 3 × 10
d) 5 × 8 ___ 2 × 20
e) 20 ÷ 10 ___ 18 ÷ 6
f) 44 ÷ 11 ___ 36 ÷ 6
g) 2 × 4 × 5 ___ 10 × 1 × 4
h) 1 + 2 + 3 ___ 3 + 2 + 1

3) Copy the calculations and put the correct symbol in the gap.
Choose from: <, > or =
a) 45 ___ 50 – 5
b) 25 ___ 2 × 20
c) 8 × 7 ___ 7 × 8
d) 80 ÷ 4 ___ 40 ÷ 2
e) $\frac{1}{2}$ of 30 ___ 7 × 2
f) $\frac{1}{4}$ of 28 ___ 20 – 12

 4) Copy the calculations and put in a number to make them true:
a) 7 + ___ = 20
b) 2 × ___ < 30
c) 40 – 23 > ___
d) 5 × 5 ≠ 4 + ___
e) 36 – 9 < ___ × 3
f) 18 ÷ 6 ≠ 30 ÷ ___

Introduction to algebra

> **Now try this!**
>
> Use any of the digits 2, 3, 9, 10 once in the calculation below to make it true.
>
> ____ × ____ < ____ × ____
>
> How many different answers can you find?
> How do you know you have found them all?

Simple equations

 I will learn to solve simple equations.

Remember, remember

= means **is equal to** (has the same value).

An **equation** is a number sentence where one side is equal to the other side.

These are all equations:

3 + 5 = 8

8 = 3 + 5

3 + 5 = 1 + 7

We can find **missing values** in equations.

This is called **solving** an equation.

Chapter 12 Equations

Example

21 = 15 + ◇

What number when added to 15 gives 21?   ◇ = 6

⌂ × 3 = 18

What number multiplied by 3 gives 18?   ⌂ = 6

♡ − 7 = 12

What number gives 12 when 7 is subtracted?   ♡ = 19

7 = ☐ ÷ 4

What number gives 7 when divided by 4?   ☐ = 28

Exercise 2

 1) What does ◇ stand for in each equation?

a) ◇ + 10 = 25   b) 18 + ◇ = 30

c) 12 = ◇ + 4   d) 18 = 12 + ◇

Introduction to algebra

2) What does ♡ stand for in each equation?
   a) ♡ − 5 = 12
   b) 14 − ♡ = 5
   c) 25 = ♡ − 20
   d) 100 = 118 − ♡

3) What does ⌂ stand for in each equation?
   a) ⌂ × 5 = 30
   b) 9 × ⌂ = 36
   c) 45 = 5 × ⌂
   d) 99 = ⌂ × 11

4) What does ☐ stand for in each equation?
   a) ☐ ÷ 10 = 12
   b) 56 ÷ ☐ = 7
   c) 11 = 66 ÷ ☐
   d) 5 = ☐ ÷ 6

5) Talk to a partner or your teacher and explain what you think Zara means by 'the opposite'.
   Do you agree with Zara?

   > I can find the missing number in an equation by doing the opposite: when there is an add, I *subtract*; when there is a *multiply*, I divide.

6) Work out what the symbol ◇ stands for in each equation:
   a) 3 × ◇ = 12
   b) ◇ + 5 = 35
   c) 20 = ◇ − 7
   d) 12 = ◇ ÷ 2
   e) 23 = 28 − ◇
   f) 3 = ◇ ÷ 8

Chapter 12 Equations

Missing operations

> 💡 I will learn to find the missing operation in an equation.

You regularly use the mathematical symbols +, –, × and ÷
In mathematics these are called **operations**.
They do something to a number.

Example

Which mathematical operation has been covered?

Choose: +, –, × or ÷

6 ◇ 3 = 9	8 ◯ 7 = 1	6 ✦ 5 = 30	24 ⬡ 4 = 6
◇ is +	◯ is –	✦ is ×	⬡ is ÷

Exercise 3

1) Which mathematical operation has been covered?
   Choose: +, –, × or ÷

   a) 6 ● 6 = 12   b) 7 ● 7 = 0

   c) 18 = 3 ● 6   d) 4 = 32 ● 8

   e) 30 = 6 ● 5   f) 17 = 20 ● 3

Introduction to algebra

2) Copy the equation with the correct operation: +, −, × or ÷

   a) 23 ___ 23 = 0

   b) 80 = 8 ___ 10

   c) 12 ___ 18 = 30

   d) 9 = 81 ___ 9

3) All the mathematical operations on a calculator have worn out. Abdul labels the keys A, B, C and D.

   He types in four calculations.

   Here are the answers:
   - 7 A 8 = 56
   - 4 B 5 = 9
   - 12 C 7 = 5
   - 20 D 4 = 5

   Which operation belongs on each key?

Forming equations

 I will learn to form an equation to solve a problem.

Sometimes it helps to **form an equation** to solve a mathematical problem.

Example

Amy buys 5 apples.
Rida gives her some more.
Amy now has 8 apples.
How many does Rida give her?

Form an equation using a symbol to represent the number you do not know.

5 + ● = 8

Solve the equation: ● = 3

Chapter 12 Equations

Exercise 4

1) Three chicks hatch in the morning.
   By evening there are 10 chicks.
   How many chicks hatch in the afternoon?
   a) Form an equation: 3 + ___ = 10
   b) Solve the equation to work out how many chicks hatch in the afternoon.

2) Bill has £9.
   He spends some money on a book.
   Bill has £4 left.
   How much does the book cost?
   a) Form an equation: 9 − ___ = 4
   b) Solve the equation to work out how much the book costs.

3) A plank of wood is 30 cm long.
   A piece is cut off.
   The piece left is 10 cm long.
   How long is the piece that is cut off?
   a) Which is the correct equation for solving this problem?

   | 10 − △ = 30 | 10 + 30 = △ | 10 × △ = 30 |
   | 30 ÷ 10 = △ | 30 − △ = 10 | △ − 10 = 30 |

   b) Solve the equation.

Introduction to algebra

4) It takes Millie 32 seconds to swim a length in the pool.

It takes Rami 45 seconds.

How much longer does Rami take?

a) Which is the correct equation for solving this problem?

45 + ♡ = 32   32 − ♡ = 45   45 − ♡ = 32

45 + 32 = ♡   45 ÷ ♡ = 32

b) Solve the equation.

5) A vase has 14 tulips in it.

Some tulips die, so only 5 are left.

Form an equation and solve it to work out how many tulips die.

6) A pet shop sells 8 mice.

There are 12 mice left.

Form an equation and solve it to work out how many mice were in the shop to begin with.

Function machines

 I will learn to use a function machine to solve equations.

A **function machine** is a diagram showing a machine that takes an **input** and applies a rule to give an **output**.
This function machine multiplies by 4 (×4).

INPUT → × 4 → OUTPUT

Chapter 12 Equations

When the input is 5:

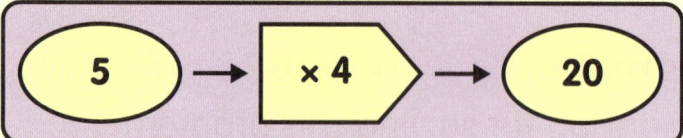

the output is 20.

Example

What is the input of this function machine?

INPUT → × 4 → 12

What multiplied by 4 gives 12?
Input = 3

Exercise 5

 1) Find the **output** of each function machine.

a) 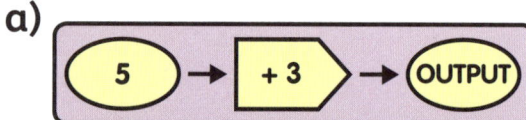 b)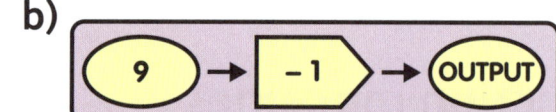

c) 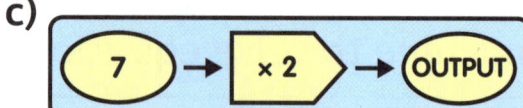 d)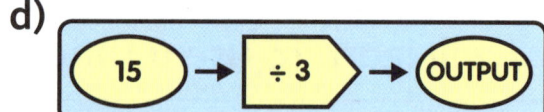

e) 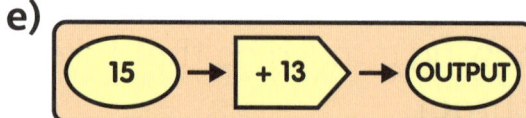 f)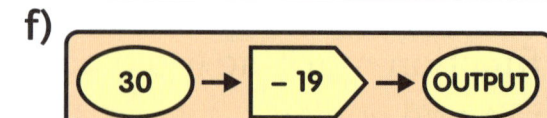

g) 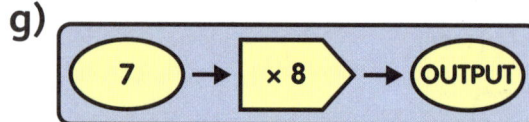 h)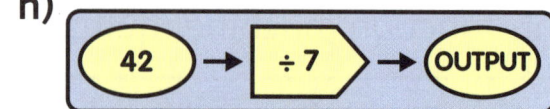

i)  j)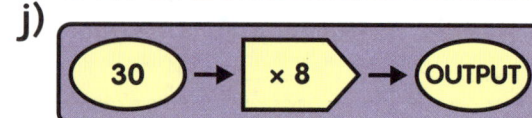

Introduction to algebra

2) Find the **input** of each function machine.

a)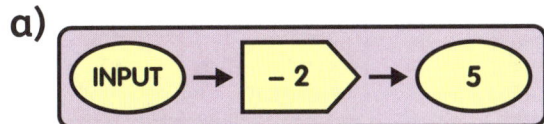

b)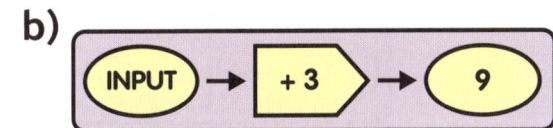

c)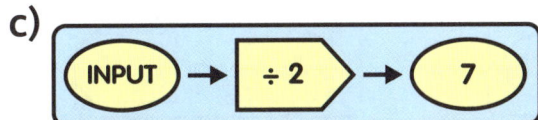

d)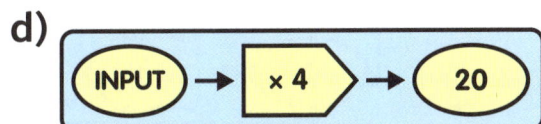

e)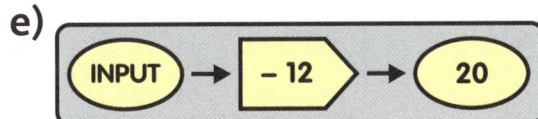

f)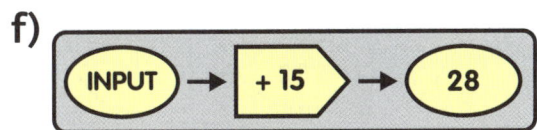

g)

h)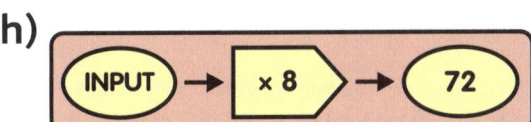

i)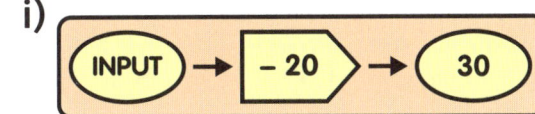

j)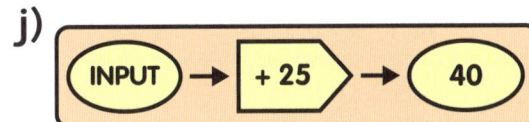

3) What is the **missing** number in each function machine?

a)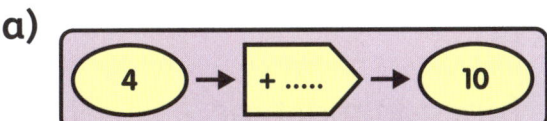

b)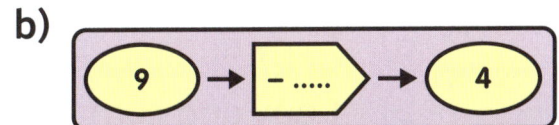

c)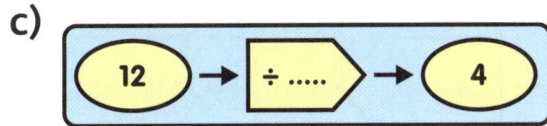

d)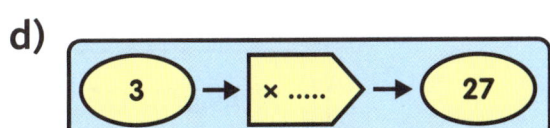

e)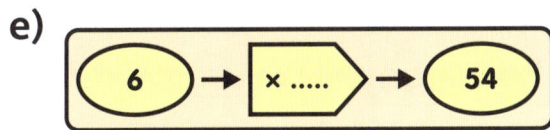

f)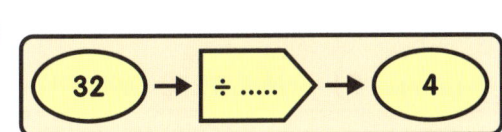

Chapter 12 Equations

4) What is the **missing** operation in each function machine?

a)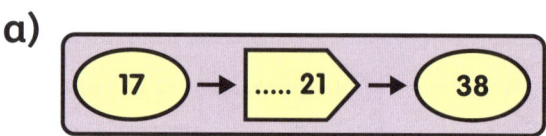

b)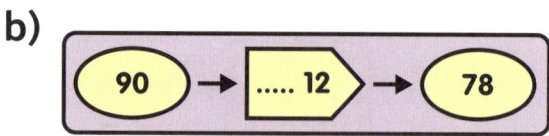

c)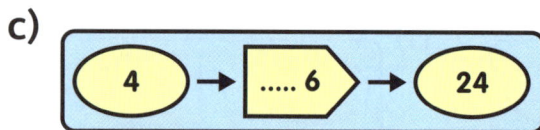

d)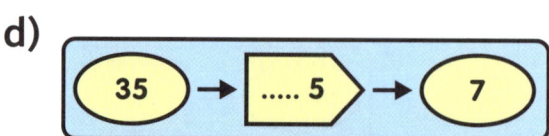

e)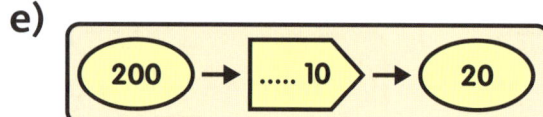

f)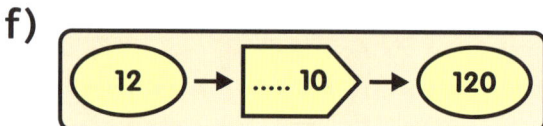

5) An SUV is a large car.

An SUV has 5 wheels (including the spare wheel).

The number of wheels for the number of SUVs is shown in the table.

Number of SUVs	1	2	3	4	5	6
Number of wheels	5	10	—	—	—	—

a) Copy and complete the table.

b) Copy and complete the function machine.

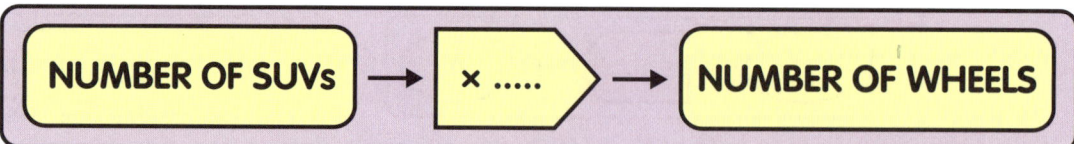

c) If there are 40 wheels, how many SUVs are there?

Introduction to algebra

Now try this!

The **inverse** in mathematics is the **opposite**; it **undoes** an **operation**.
The inverse of the function machine:

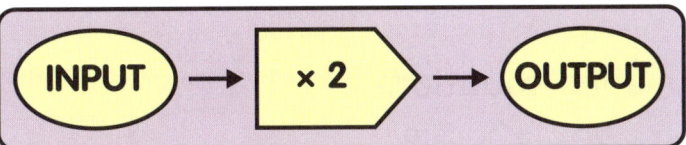

is

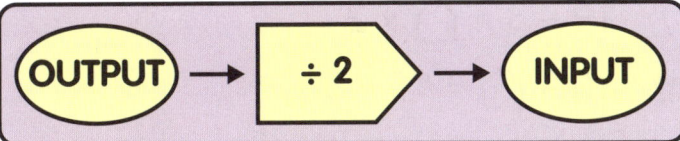

Draw inverse function machines for the function machines in question 2 on page 137.

Explain to your teacher how they might help you to answer question 2.

Revisit, review, revise

1) Match the calculations in pairs using the = sign.

   7 × 3   40 ÷ 4   39 − 19

   2 × 10   1 + 2 + 3 + 4   15 + 6

2) Copy each pair of calculations and put = or ≠ between them.
   a) 6 × 6 ___ 9 × 4
   b) 28 ÷ 7 ___ 30 ÷ 5
   c) $\frac{1}{2}$ of 22 ___ $\frac{1}{3}$ of 27

Chapter 12 Equations

3) Copy each pair of calculations and put < or > between them.
   a) 6 × 7 ___ 7 × 5
   b) $\frac{1}{2}$ of 50 ___ 3 × 8
   c) 16 − 12 ___ 12 + 4

4) What does the ⌂ stand for in each equation?
   a) ⌂ + 9 = 15
   b) 20 − ⌂ = 8
   c) 32 = ⌂ × 4
   d) 4 = 16 ÷ ⌂
   e) ⌂ + 5 = 2 × 4
   f) 19 − 3 = ⌂ × 2
   g) 20 ÷ 5 = 8 ÷ ⌂
   h) 20 − ⌂ = 10 + 5

5) Which operation does ☐ stand for in each equation?
   a) 12 ☐ 3 = 36
   b) 12 ☐ 3 = 4
   c) 12 ☐ 3 = 15
   d) 12 ☐ 3 = 9
   e) 17 = 33 ☐ 16
   f) 9 = 63 ☐ 7
   g) 45 = 5 ☐ 9
   h) 45 = 29 ☐ 16

6) Emma has £18.
   She buys a sandwich.
   She now has £13 left.
   a) Which equation would you solve to find the cost of the sandwich?

   13 + 18 = ◇   18 × ◇ = 13   18 + ◇ = 13
   18 − ◇ = 13   18 + 13 = ◇

   b) Solve the equation.

Introduction to algebra

7) What is the **output** of each function machine?

a)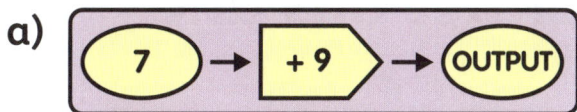

b)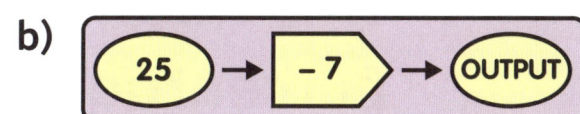

c)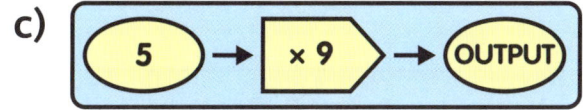

d)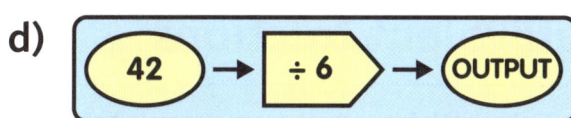

8) What is the **input** of each function machine?

a)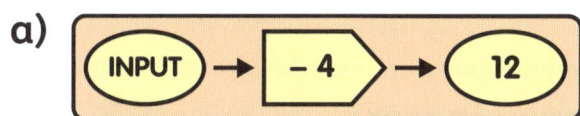

b)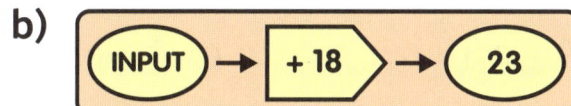

c)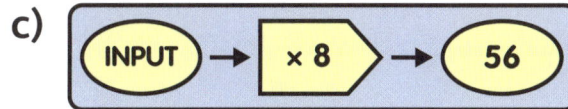

d)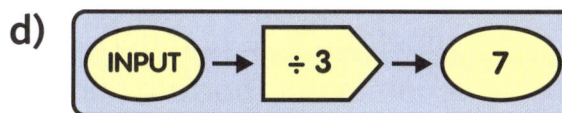

9) What number is **missing** in each function machine?

a)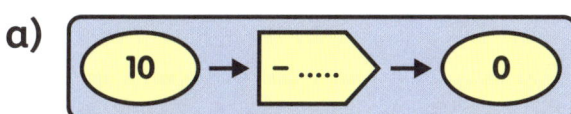

b)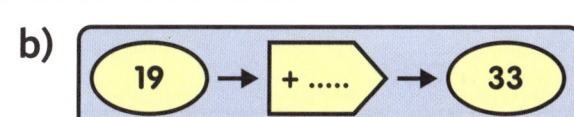

c)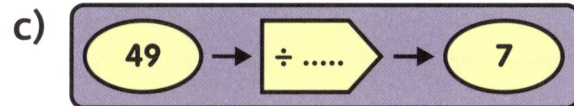

d)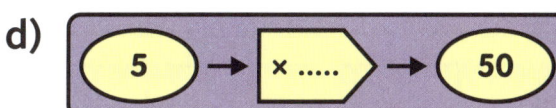

13 Fractions and percentages
Understanding fractions and percentages

Equivalent fractions

 I will learn to recognise and find equivalent fractions.

A **fraction** is part of a whole.

A fraction has a **numerator** and a **denominator**.

$\dfrac{1}{3}$ ← numerator (top number)
← denominator (bottom number)

Fractions can have **different numerators** and **different denominators**, but represent the **same** amount.

For example:

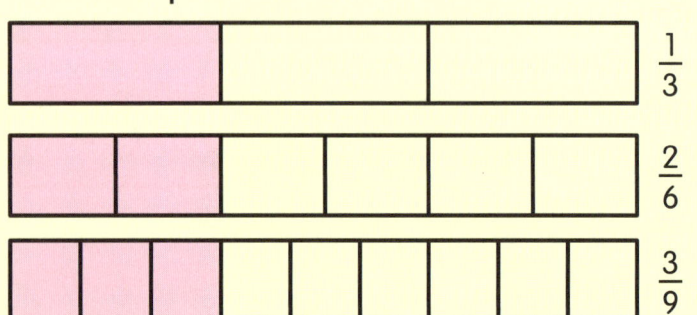

$\dfrac{1}{3} = \dfrac{2}{6} = \dfrac{3}{9}$ These are called **equivalent fractions**.

You can find equivalent fractions by multiplying the numerator and denominator by the same number.

For example:

$\dfrac{1}{3} = \dfrac{1 \times 2}{3 \times 2} = \dfrac{2}{6}$ $\dfrac{1}{3} = \dfrac{1 \times 3}{3 \times 3} = \dfrac{3}{9}$

Understanding fractions and percentages

Exercise 1

 1) Copy and complete these equivalent fractions:

a) $\dfrac{1}{2} = \dfrac{1 \times 2}{2 \times 2} = \dfrac{2}{\square}$

b) $\dfrac{2}{3} = \dfrac{2 \times 2}{3 \times 2} = \dfrac{4}{\square}$

c) $\dfrac{3}{4} = \dfrac{3 \times 6}{4 \times 6} = \dfrac{18}{\square}$

d) $\dfrac{1}{10} = \dfrac{1 \times 6}{10 \times 6} = \dfrac{\square}{\square}$

e) $\dfrac{3}{5} = \dfrac{3 \times 3}{5 \times 3} = \dfrac{\square}{\square}$

f) $\dfrac{7}{10} = \dfrac{7 \times 4}{10 \times 4} = \dfrac{\square}{\square}$

2) In your jotter, write an equivalent fraction to each fraction by multiplying the numerator and denominator by 5:

a) $\dfrac{1}{2}$ b) $\dfrac{2}{3}$ c) $\dfrac{5}{8}$ d) $\dfrac{9}{10}$

3) In your jotter, write an equivalent fraction to each fraction by multiplying the numerator and denominator by 9:

a) $\dfrac{1}{4}$ b) $\dfrac{2}{3}$ c) $\dfrac{5}{8}$ d) $\dfrac{6}{7}$

4) Copy and complete these equivalent fractions with denominator 100:

a) $\dfrac{10 \times 4}{25 \times 4} = \dfrac{\square}{100}$

b) $\dfrac{8}{25} = \dfrac{\square}{100}$

c) $\dfrac{9 \times 5}{20 \times 5} = \dfrac{\square}{100}$

d) $\dfrac{1}{20} = \dfrac{\square}{100}$

e) $\dfrac{1 \times 25}{4 \times 25} = \dfrac{\square}{100}$

f) $\dfrac{3}{4} = \dfrac{\square}{100}$

Chapter 13 Fractions and percentages

5) Which fractions are equivalent to $\frac{1}{3}$?

| $\frac{6}{9}$ | $\frac{4}{12}$ | $\frac{5}{15}$ | $\frac{6}{8}$ | $\frac{10}{30}$ |

 6) Ann writes an equivalent fraction to $\frac{7}{9}$

She says that the denominator is 63.

What is the **numerator**?

Simplifying fractions

 I will learn to simplify fractions.

You can **simplify fractions** by **dividing the numerator and denominator** by the same number.

For example:

$$\frac{4 \div 2}{6 \div 2} = \frac{2}{3}$$

When there is no other number that can divide the numerator and denominator, then the fraction cannot be simplified any further. It is written in its **simplest form**.

For example:

$$\frac{4 \div 2}{6 \div 2} = \frac{2}{3}$$

It is not possible to divide 2 and 3 by the same number again.

$\frac{4}{6}$ in its **simplest form** is $\frac{2}{3}$

Understanding fractions and percentages

Sometimes it is possible to further simplify a fraction.
For example:

$$\frac{12 \div 2}{18 \div 2} = \frac{6}{9}$$ — It is possible to divide 6 and 9 by 3.

$$\frac{6 \div 3}{9 \div 3} = \frac{2}{3}$$ — $\frac{12}{18}$ in its **simplest form** is $\frac{2}{3}$

You can also simplify $\frac{12}{18}$ in one step. The answer is the same.

$$\frac{12 \div 6}{18 \div 6} = \frac{2}{3}$$

Exercise 2

1) Copy and complete to simplify these fractions:

   a) $\frac{6 \div 2}{8 \div 2} = \frac{3}{\square}$

   b) $\frac{12 \div 3}{30 \div 3} = \frac{4}{\square}$

   c) $\frac{20 \div 10}{50 \div 10} = \frac{\square}{5}$

   d) $\frac{12 \div 4}{20 \div 4} = \frac{\square}{\square}$

   e) $\frac{18 \div 6}{30 \div 6} = \frac{\square}{\square}$

   f) $\frac{24 \div 8}{32 \div 8} = \frac{\square}{\square}$

2) Simplify these fractions by dividing the numerator and denominator by 5:

   a) $\frac{15}{20}$ b) $\frac{10}{25}$ c) $\frac{15}{40}$ d) $\frac{20}{35}$ e) $\frac{45}{50}$

Chapter 13 Fractions and percentages

3) Which fractions simplify to $\frac{1}{4}$?

| $\frac{2}{8}$ | $\frac{4}{16}$ | $\frac{5}{25}$ | $\frac{4}{40}$ | $\frac{7}{28}$ |

4) Tanvi simplifies $\frac{18}{24}$ like this: $\frac{18}{24} = \frac{9}{12}$

   a) Has Tanvi written $\frac{18}{24}$ in its simplest form? Yes or no.

   b) If no, write the fraction in its simplest form in your jotter.

5) Are these fractions written in their simplest form? Yes or no.

   a) $\frac{2}{4}$ b) $\frac{6}{9}$ c) $\frac{5}{6}$ d) $\frac{8}{24}$ e) $\frac{7}{12}$

6) In your jotter, write each fraction in its simplest form:

   a) $\frac{5}{30}$ b) $\frac{12}{16}$ c) $\frac{15}{18}$ d) $\frac{14}{35}$
   e) $\frac{9}{27}$ f) $\frac{12}{24}$ g) $\frac{24}{40}$ h) $\frac{45}{81}$

7) Alec is thinking of a fraction.
   He says in its simplest form his fraction is $\frac{6}{7}$
   In your jotter, write two possible fractions that Alec may have simplified.

Understanding fractions and percentages

What is a percentage?

> 💡 I will learn that percentages are parts of 100.

When a shape or amount is divided into **100 equal parts**, then each part is a **percent**.

The symbol for percent is %.

1% = 1 part out of 100 equal parts

For example:

This large square is divided into 100 equal parts.

38 parts out of 100 equal parts are purple.

38% is purple.

6 parts out of 100 equal parts are pink.

6% is pink.

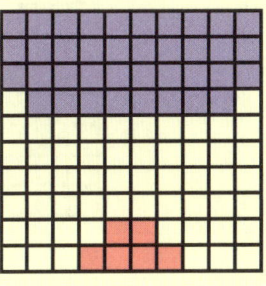

Exercise 3

1) What percentage of each large square is coloured? Write your answer as ____ %.

   a) 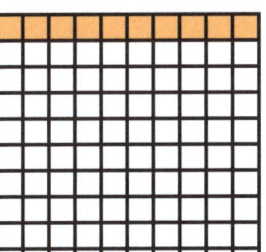   b) 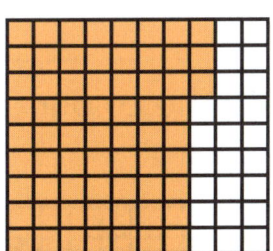   c)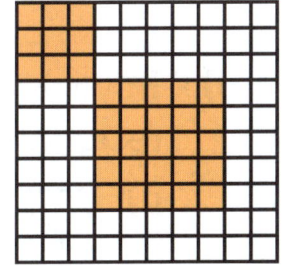

2) In question 1, what percentage of each large square is **not** coloured?

Chapter 13 Fractions and percentages

3) In your jotter, write the colours as percentages of each large square.

a)

blue = ____ %

red = ____ %

b)

yellow = ____ %

green = ____ %

4) In question 3, what percentage of each large square is **not** coloured?

5) Without counting the white squares, write a sentence in your jotter to explain how you could answer question 4.

6) In your jotter, write the colour as a percentage of each rectangle.

a)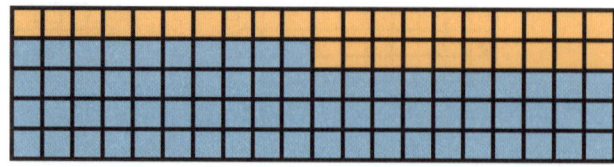

orange = ____ %

blue = ____ %

b)

yellow = ____ %

grey = ____ %

Understanding fractions and percentages

7) In every 100 people, 10 are left-handed.
What percentage are left-handed?

 8) In every 100 children in Scotland:
- 39 are driven to school
- 44 walk to school
- 4 cycle to school.
a) What percentage are driven to school?
b) What percentage walk to school?
c) What percentage cycle to school?
d) What percentage travel to school a different way?

Percentages and fractions

💡 I will learn about equivalent percentages and fractions.

You can write a **percentage as a fraction** with a **denominator of 100**.

$1\% = \dfrac{1}{100}$ = 1 part out of 100 equal parts.

For example:

$23\% = \dfrac{23}{100}$ $9\% = \dfrac{9}{100}$

Sometimes you may need to simplify the fraction.

For example:

$5\% = \dfrac{5 \div 5}{100 \div 5} = \dfrac{1}{20}$ $24\% = \dfrac{24 \div 4}{100 \div 4} = \dfrac{6}{25}$

Chapter 13 Fractions and percentages

You can write a fraction with a **denominator of 100** as a percentage.
For example:

$\dfrac{39}{100} = 39\%$ $\dfrac{7}{100} = 7\%$

Sometimes you may need to write an **equivalent fraction** with a **denominator of 100**.
For example:

$\dfrac{1 \times 50}{2 \times 50} = \dfrac{50}{100} = 50\%$ $\dfrac{9 \times 4}{25 \times 4} = \dfrac{36}{100} = 36\%$

Exercise 4

 1) Copy and complete to write these percentages as fractions:

a) $37\% = \dfrac{\square}{100}$ b) $11\% = \dfrac{\square}{100}$ c) $99\% = \dfrac{\square}{100}$

 2) Copy and complete to write these fractions as percentages:

a) $\dfrac{43}{100} = \underline{}\%$ b) $\dfrac{69}{100} = \underline{}\%$ c) $\dfrac{19}{100} = \underline{}\%$

3) In your jotter, write the parts of each large square that are coloured as a percentage and a fraction.

a) $\dfrac{\square}{100} = \underline{}\%$ b) $\dfrac{\square}{100} = \underline{}\%$

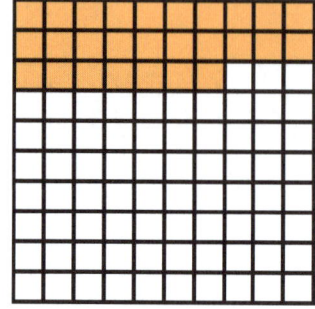   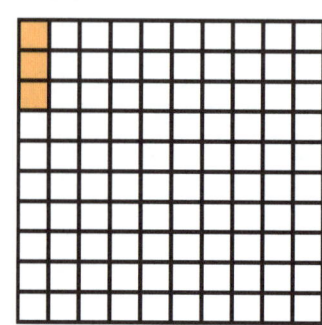

Understanding fractions and percentages

4) In your jotter, write each percentage as a fraction.
   a) 57%   b) 89%   c) 21%   d) 77%

5) Copy and complete to write each percentage as a fraction in its simplest form:
   a) $12\% = \dfrac{12 \div 4}{100 \div 4} = \dfrac{\square}{25}$
   b) $70\% = \dfrac{70 \div 10}{100 \div 10} = \dfrac{\square}{\square}$

6) In your jotter, write each percentage as a fraction in its simplest form.
   a) 10%   b) 90%   c) 18%   d) 6%

7) Copy and complete to write each fraction with a denominator of 100 and then as a percentage:
   a) $\dfrac{7 \times 5}{20 \times 5} = \dfrac{\square}{100} = \underline{}\%$
   b) $\dfrac{19 \times 2}{50 \times 2} = \dfrac{38}{\square} = \underline{}\%$

☀ 8) In your jotter, write each fraction with a denominator of 100 and then as a percentage.
   a) $\dfrac{3}{20} = \dfrac{\square}{100} = \underline{}\%$
   b) $\dfrac{9}{50} = \dfrac{\square}{100} = \underline{}\%$
   c) $\dfrac{3}{10} = \dfrac{\square}{100} = \underline{}\%$
   d) $\dfrac{7}{25} = \dfrac{\square}{100} = \underline{}\%$

Chapter 13 Fractions and percentages

Now try this!

Fraction percentage pairs

A game for four players.

Each player cuts out four cards.

On one card, write a percentage from 1 to 100.

On another card, write the percentage as a fraction.

Repeat on the two other cards with a different percentage and its equivalent fraction.

Playing the game

- Put all 16 cards face down on the table, so you cannot see what is on them.
- Move the cards around.
- Take it in turns to turn over any two cards.
- If the percentage and fraction match, you pick up the two cards.
- If you turn over two percentages, two fractions or a percentage and fraction that are not equivalent, you turn the cards face down again.
- When all 16 cards are gone, the person with the most cards is the winner.

Revisit, review, revise

1) What percentage of the large square is:
    a) coloured
    b) **not** coloured?

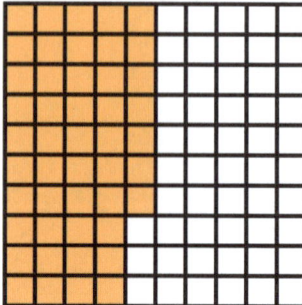

Understanding fractions and percentages

2) In your jotter, write each percentage as a fraction.
   a) 29%
   b) 53%
   c) 71%
   d) 97%

3) Copy and complete to work out the equivalent fraction:
   a) $\dfrac{3 \times 8}{4 \times 8} = \dfrac{\square}{\square}$
   b) $\dfrac{5 \times 2}{9 \times 2} = \dfrac{\square}{\square}$

4) Which pairs of fractions are equivalent?

$\dfrac{1}{2}$	$\dfrac{4}{24}$	$\dfrac{6}{12}$	$\dfrac{24}{30}$	$\dfrac{1}{6}$	$\dfrac{4}{5}$

5) In your jotter, write each fraction in its simplest form.
   a) $\dfrac{6}{10}$
   b) $\dfrac{9}{30}$
   c) $\dfrac{8}{24}$
   d) $\dfrac{9}{36}$
   e) $\dfrac{8}{64}$

6) In your jotter, write each percentage as a fraction in its simplest form.
   a) 1%
   b) 20%
   c) 25%
   d) 66%

7) For every 100 people in Scotland, 17% are **younger** than 16 years old.
   a) What percentage of people in Scotland are **older** than 16 years old?
   b) In your jotter, write your answer to part a) as a fraction.

8) In your jotter, write each fraction with a denominator of 100 and then as a percentage.
   a) $\dfrac{1}{10}$
   b) $\dfrac{7}{20}$
   c) $\dfrac{9}{25}$
   d) $\dfrac{1}{50}$
   e) $\dfrac{3}{5}$

14 Perimeter and area
Calculating perimeter and area

Perimeter and area: squares and rectangles

💡 I will learn to calculate the perimeter and area of squares and rectangles.

The **perimeter** of a shape is the total **distance** around the outside.
Perimeter is measured in units of length, for example: **metres**, **centimetres**, **millimetres**.

Example

Work out the perimeter of the rectangle:
Opposite sides of a rectangle are the same length.
Perimeter of rectangle = 11 + 4 + 11 + 4 = 30 cm

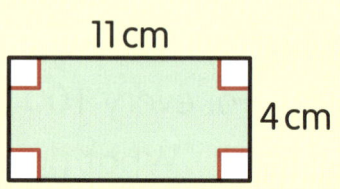

The **area** of a shape is the space it takes up.
Area is measured in units squared, for example: **cm²**, **m²**, **mm²**.
Area of a rectangle = length × height

Example

Work out the area of the square:
All the sides of a square are equal.
Area of square = 5 × 5 = 25 cm²

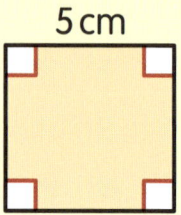

Calculating perimeter and area

Exercise 1

  1) a) Copy the rectangle in your jotter and write on the missing lengths.

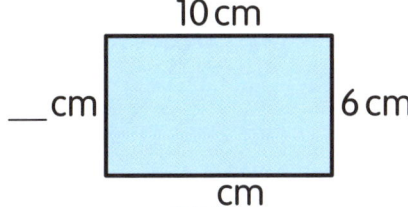

b) Copy and complete:

perimeter = 10 + 6 + ____ + ____ = ____ cm

area = 10 × ____ = ____ cm²

  2) Alin and Faizal are working out the **perimeter** of a square.

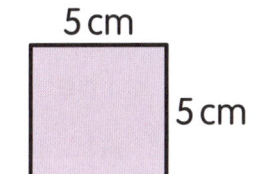

Alin says that the perimeter is 4 × 5 = 20 cm.

Faizal says that the perimeter is 5 + 5 + 5 + 5 = 20 cm.

a) Whose method do you prefer?

b) Work out the **perimeter** of a square with sides:

i) 10 cm   ii) 7 cm   iii) 9 cm

3) Calculate the
a) perimeter   b) area

of these rectangles and squares.

i)   ii)   iii)

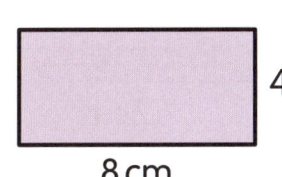   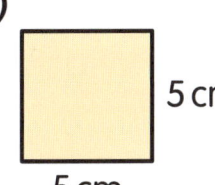   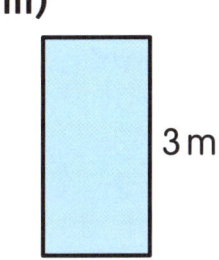

Chapter 14 Perimeter and area

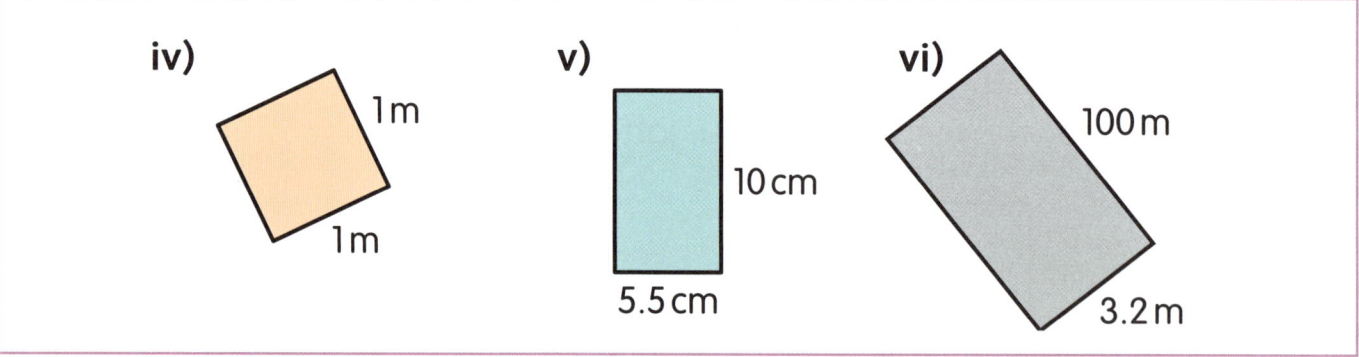

iv) 1 m, 1 m

v) 10 cm, 5.5 cm

vi) 100 m, 3.2 m

4) A square has perimeter 12 cm.
   a) What is the length of one side?
   b) Work out the area of the square.

5) A farmer puts 100 m of hedge around a field. The field is 41 m long.

   __ m
   41 m

   a) How wide is the field?
   b) What is the area of the field?

 6) A square has area 36 cm². What is the perimeter?

 7) A rectangle has area 20 cm².

   The lengths of the sides are whole numbers.

   Copy and complete the table to show all the possible lengths, widths and perimeters for this rectangle.

Length	Width	Perimeter
1	20	42 cm
2	10	__ cm
__	__	__ cm

Calculating perimeter and area

 8) A carpenter makes a rectangular wooden picture frame 20 cm by 15 cm.

Strips of wood can be bought in half-metre lengths.

The prices for the materials are:
- wood: £3.50 per half-metre length
- glass: £7.25 per 100 cm².

How much does the carpenter spend on materials?

Now try this!

A farmer has 100 m of fence.

What is the largest rectangular area she can fence off?

Draw a table to show your results:

Length	Width	Area
____ m	____ m	____ m²
____ m	____ m	____ m²

What about if she has 200 m of fence?

What about 400 m?

Explain what you notice.

Revisit, review, revise

1) A rectangle has sides of length 7 m and 4 m.

Work out the:

a) perimeter   b) area.

Chapter 14 Perimeter and area

2) A square has sides 8 cm long.
   What is the:
   a) perimeter
   b) area?

3) A rectangular rabbit enclosure is made using 10 m of fence. It is 2 m wide.
   What is the area of the enclosure?

4) Work out the perimeter of this rectangle:

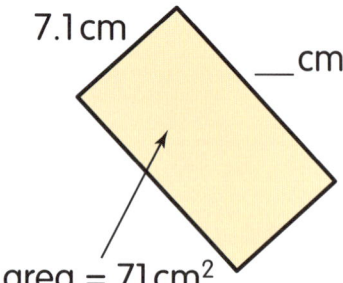

7.1 cm

___ cm

area = 71 cm^2

15 3D objects and volume
2D representation of 3D shapes

Volume

 I will learn to find the volume of a cuboid.

Remember, remember

Volume is a measurement of how much space is taken up by a 3-dimensional (3D) shape.

Volume can be measured in **cubic centimetres (cm³)**.

This cube measures 1 cm by 1 cm by 1 cm.

It has a volume of 1 cm³.

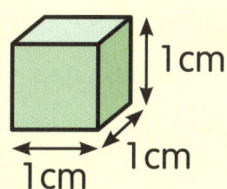

Volume can also be measured in **cubic metres (m³)**.

This cube measures 1 m by 1 m by 1 m.

It has a volume of 1 m³.

You can calculate the volume of a cuboid by working out how many cm³ or m³ it contains.

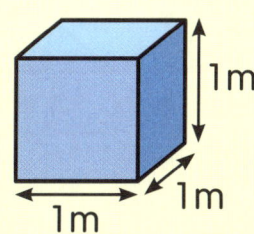

Example

This cuboid is made of 3 × 2 × 4 centimetre cubes.

Volume = 3 × 2 × 4 = 24 cm³

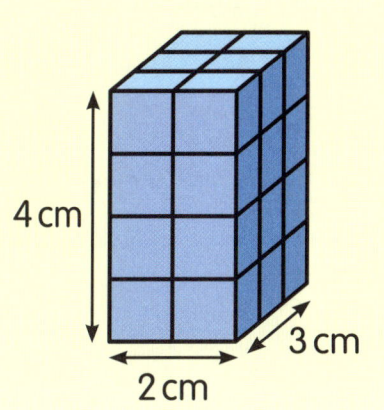

Chapter 15 3D objects and volume

Volume of a cuboid = length × width × height

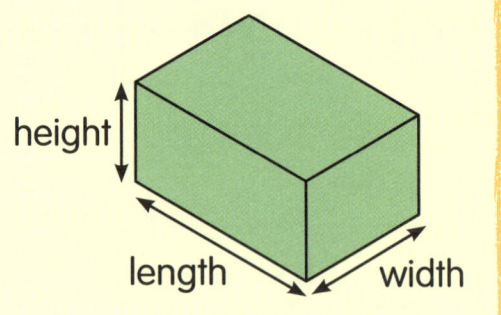

Exercise 1

1) These shapes are made from 1 cm cubes.
   Calculate the volume of each shape.

   a) 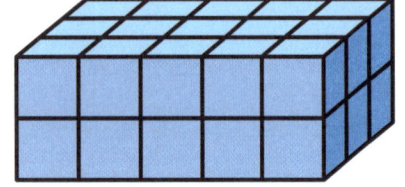   b)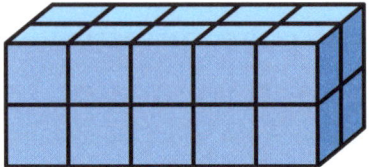

   c)    d)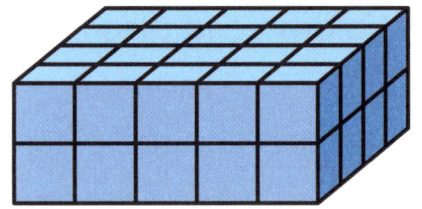

2) These shapes are cuboids made from 1 cm cubes.
   Calculate the volume of each cuboid.

   a)    b)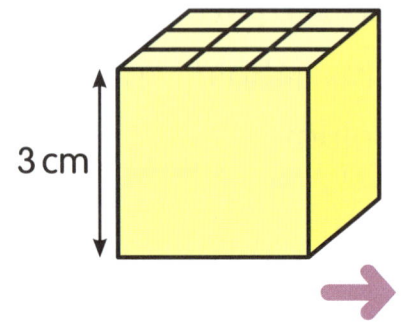

2D representation of 3D shapes

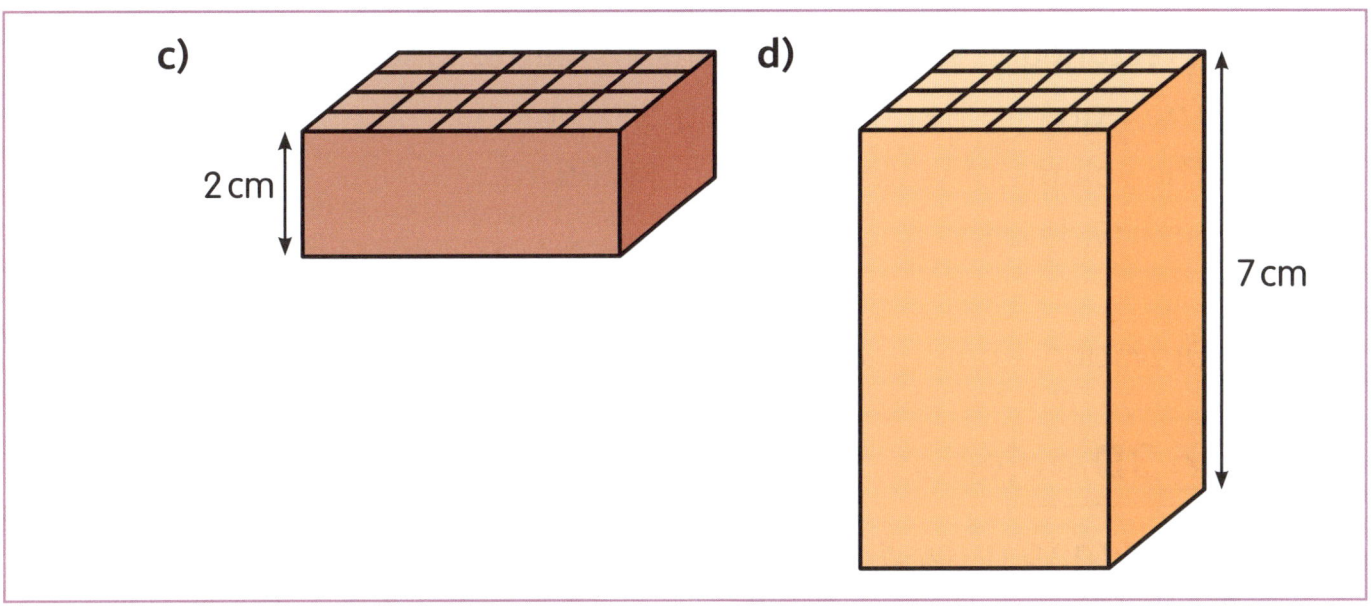

3) Calculate the volume of each cuboid.

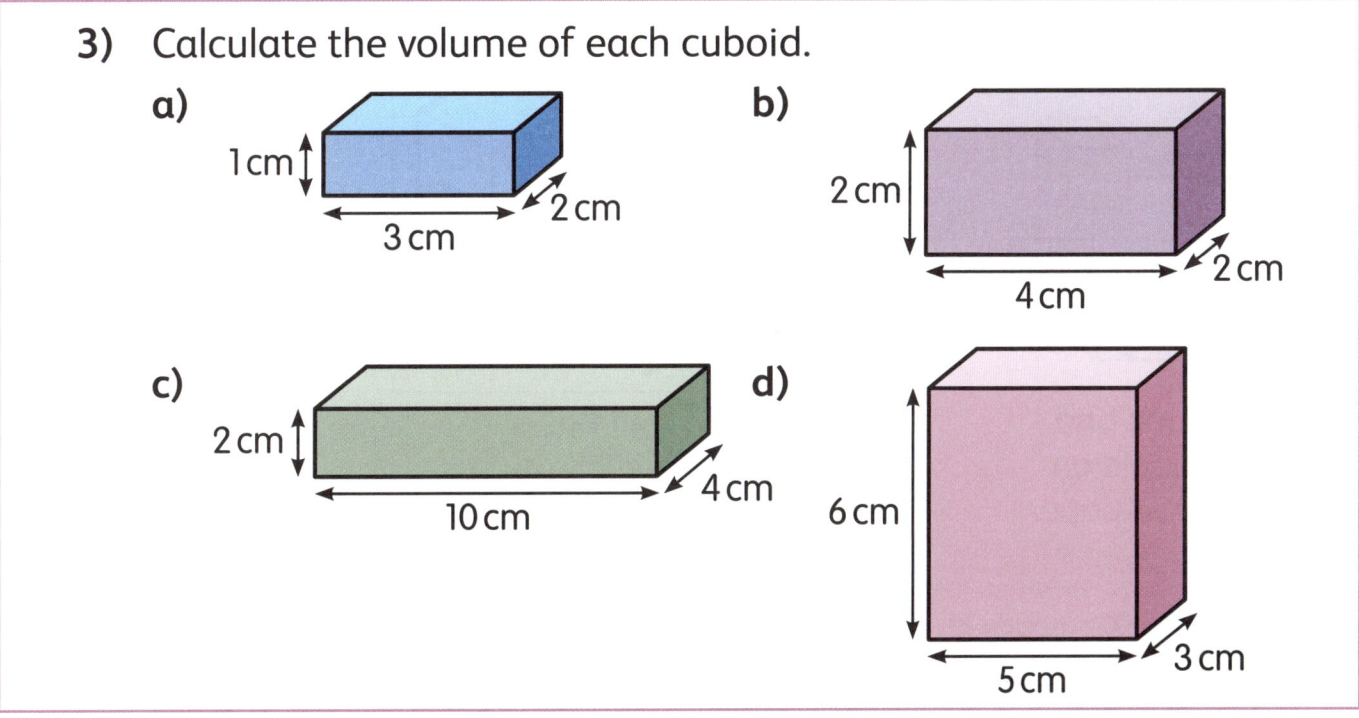

4) This cardboard box has volume 400 cm³.
What is the height of the box?

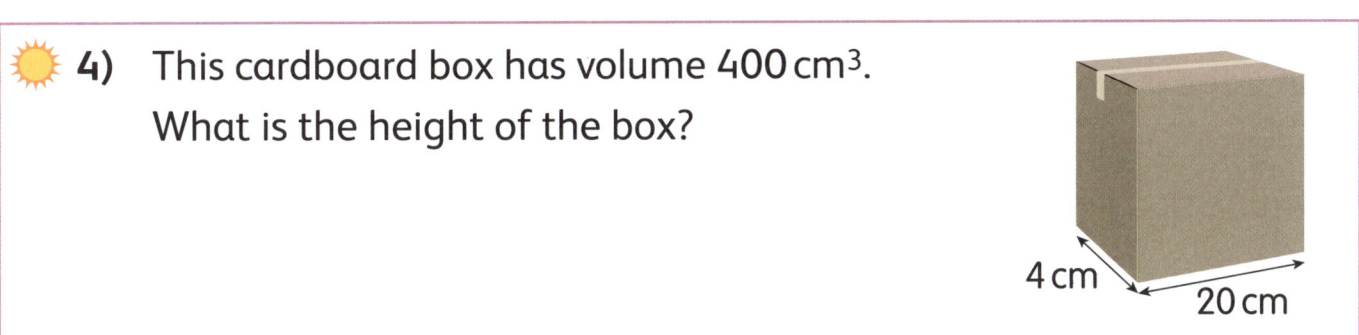

Chapter 15 3D objects and volume

 5) A rectangular swimming pool is 12 metres long.
   The water in the pool is 2 metres deep.
   The volume of water in the pool is 96 m³.
   How wide is the pool?

 6) Draw and label three different cuboids with volume 24 cm³.

Drawing 3D objects

 I will learn to draw 3D shapes on squared and isometric paper.

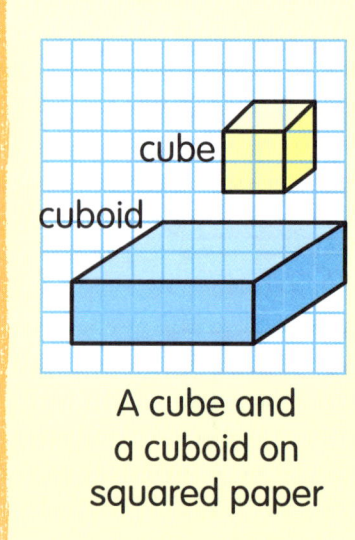

A cube and a cuboid on squared paper

Cubes and cuboids on isometric paper

Remember, remember

The flat surfaces of 3D shapes are called **faces**.
An **edge** is where two faces meet.
A **vertex** is where edges meet.
A cube has 6 square faces, 12 edges and 8 vertices.

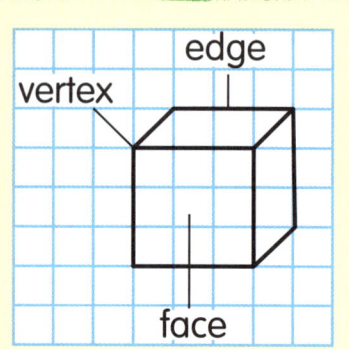

2D representation of 3D shapes

Exercise 2

You will need squared paper and triangular dotty isometric paper.

1) a) On squared paper draw a square.

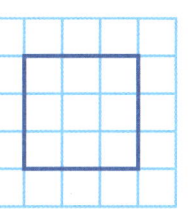

   b) Draw 3 parallel lines from three vertices.

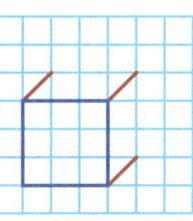

   c) Join the lines to draw a cube.

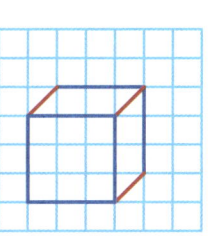

2) Draw a cuboid on squared paper.

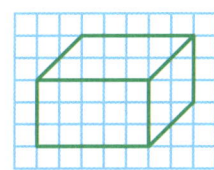

3) a) On isometric paper, draw three edges of a cuboid.

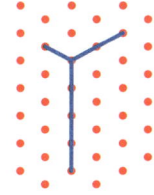   Make sure you line up the dotty paper the correct way.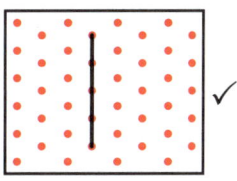

   b) Draw the other edges of the cuboid.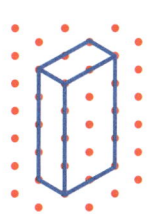

Chapter 15 3D objects and volume

4) On isometric paper, copy and complete these drawings of cuboids.

a)   b) 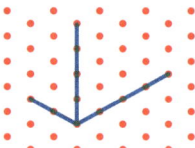   c)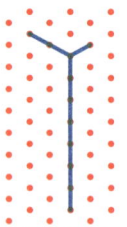

 5) a) Use isometric paper to draw a cube with edges 4 cm long. Use a ruler to measure the sides.
 b) How many faces of the cube can you see?
 c) How many edges can you see?
 d) How many vertices can you see?

 6) Draw and colour this shape on isometric paper.

Now try this!

Here is an arrangement of 4 cubes:

You can draw this shape on isometric paper:

How many other arrangements of four cubes can you draw?
Compare your answers with a partner or the class.

2D representation of 3D shapes

Nets of cubes

 I will learn to recognise and draw nets of cubes.

A **net** of a **3D shape** is what it would look like if it was opened out and laid flat.

Here is a **cube** with edges 1 cm long: One net of the cube is:

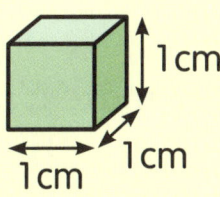
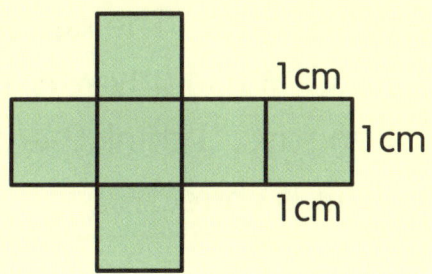

Exercise 3

You will need squared paper and a ruler for this exercise.

1) Look at the cube in the yellow box.
   a) What shape are the faces of the cube?
   Look at the net of the cube.
   b) What shapes is the net made from?

2) a) On squared paper, draw a full size net for a cube with sides 3 cm long.
   b) Cut it out and fold it to make a cube.

3) a) Copy each shape onto squared paper.
   i) 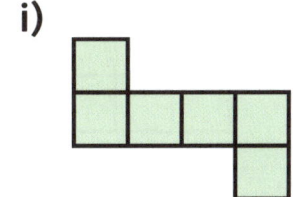   ii) 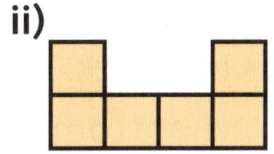   iii)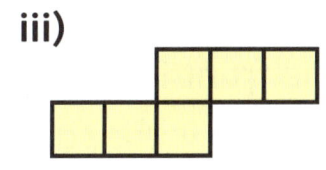

Chapter 15 3D objects and volume

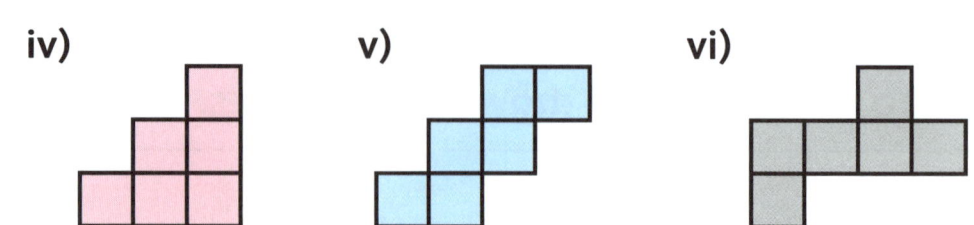

b) Cut them out and try to fold them to make cubes.
c) Which are nets of cubes?

4) Amna says: 'The net of a cube is made with 6 squares; you can join them in any way you like.'
Ben says: 'I think there are only some ways that will make a net.'
Who is correct?

5) Jen is making the net of a cube.
She draws four squares.
a) Copy this onto squared paper.
b) Draw two more squares on it to make the net of a cube.
c) How many different ways can you find to do this?

Nets of cuboids

💡 I will learn to recognise and draw nets of cuboids.

The **net** of a **cuboid** is the shape you obtain when a cuboid is laid out flat.

A cuboid has 6 faces; opposite faces are the same shape.

- The **front** is the same as the **back**.
- The **top** is the same as the **bottom**.
- The **right side** is the same as the **left side**.

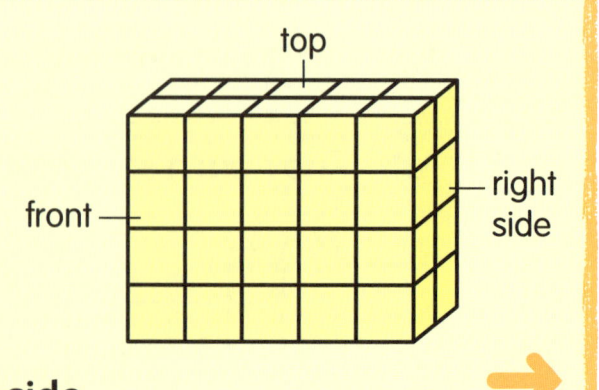

2D representation of 3D shapes

To draw the net of a cuboid:

Step 1: Draw the front and the top. Make sure the lengths of the edges are correct.

Step 2: Complete a chain of 4 rectangles by adding the back and base.

Step 3: Now add on the sides.

Exercise 4

You will need squared paper and a ruler for this exercise.

1) Here is part of a net of the cuboid.
   a) Copy this onto squared paper and add the back and base.
   b) Add the sides.
   c) Cut out the net and fold it to make a cuboid.

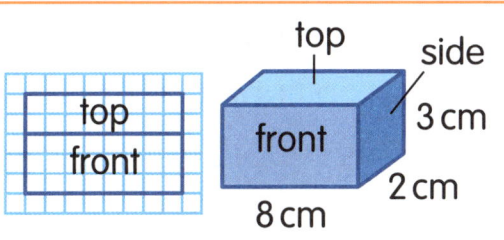

Chapter 15 3D objects and volume

2) a) On squared paper, draw a net for each cuboid.

i) 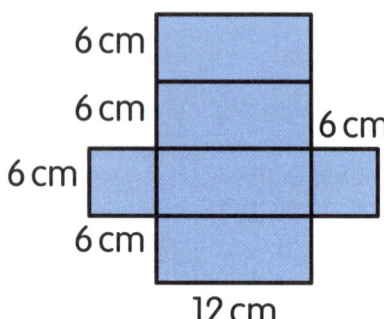   ii)   iii)

b) Cut out the nets and fold them to make cuboids.

☀ 3) On isometric paper, draw the cuboid that you can make from each net.

a)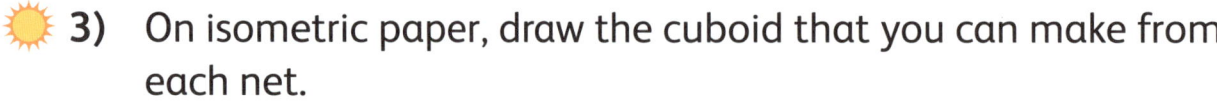

b)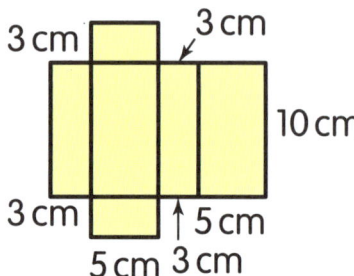

4) Work out the volume of each cuboid in question 3.

Revisit, review, revise

1) Work out the volume of:
   a) a cube with sides 4 cm
   b) a cuboid with sides 2 cm, 4 cm and 3 cm
   c) a cuboid with sides 3 m, 10 m, 4 m.

2) This cuboid has volume 80 cm³.
   What is the height?

___ cm

5 cm   2 cm

2D representation of 3D shapes

3) On isometric paper complete this drawing of a cube.

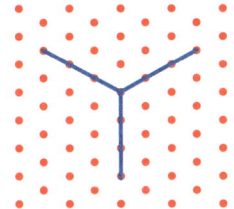

4) On squared paper complete this drawing of a cuboid.

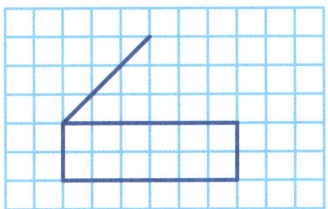

5) What 3D shape will each net make?

a)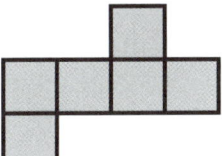

b)

16 Statistics
Understanding graphs and charts

Interpreting graphs and charts

 I will learn to represent data in graphs and charts.

Remember, remember

Data can be represented in different ways.

A **frequency table**:

Eye colour	Tally	Total
Blue	ⵊⵊ	5
Brown	|||	3
Green	||||	4
Grey			1
Hazel	||	2

← 5 children have blue eyes

A **pictograph**:

Understanding graphs and charts

A **bar chart**:

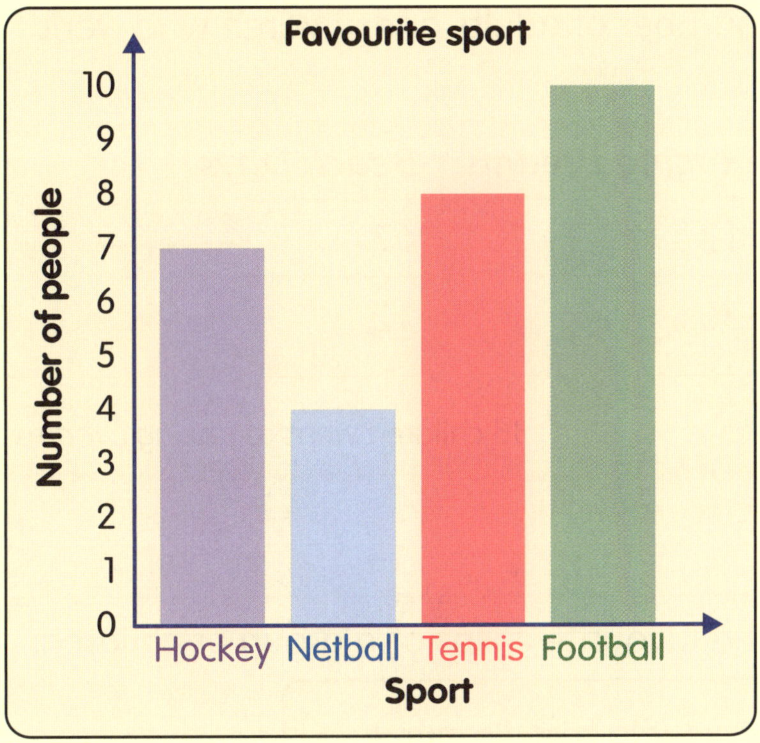

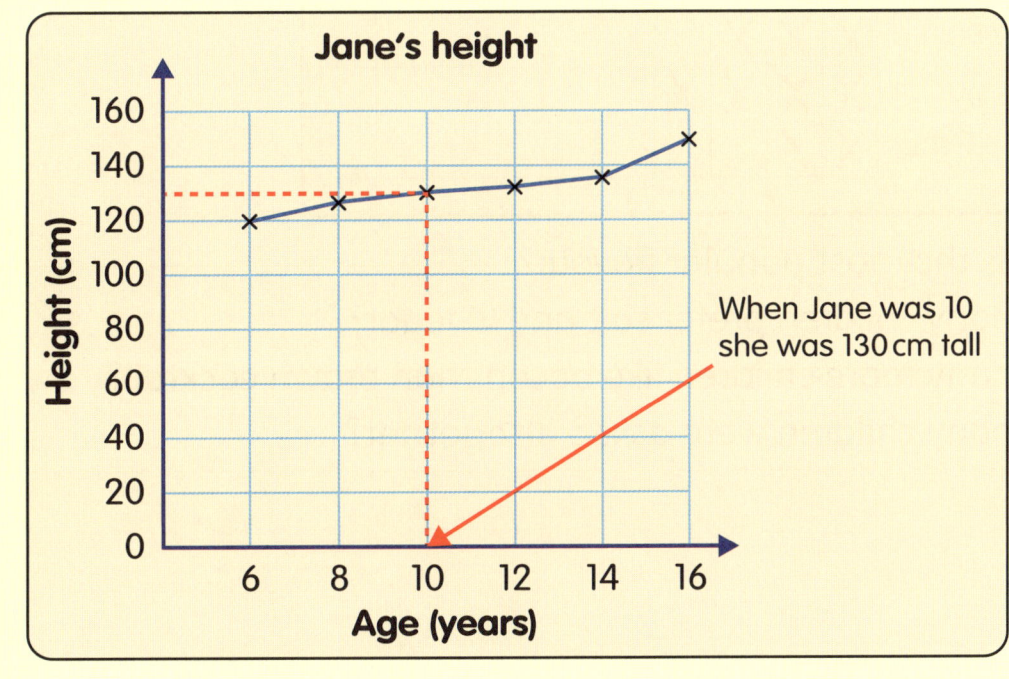

Line graphs can be used to compare values which change with time.
This line graph shows Jane's height from age 6 to 16.

When Jane was 10 she was 130 cm tall

Chapter 16 Statistics

Two-way tables show information with two characteristics.

This two-way table shows the number of adults and children who went to different films at a cinema:

	Happy Toes	Racing Dreams	Beach Days
adults	25	13	47
children	42	36	12

36 children went to *Racing Dreams*

Exercise 1

1) This pictograph shows the favourite crisps of a group of children:

    Key: X = 4 people   **Favourite crisp flavour**

Flavour								
Salt and vinegar	X	X	X	X				
Cheese and onion	X	X	X	X	X	X		
Prawn cocktail	X	X	X	`				
Ready salted	X	X	Y					
Bacon	X	X	X	X	`			
Other	X	Y						

    a) What is the most popular flavour?
    b) How many children prefer salt and vinegar?
    c) How many more children like bacon than prawn cocktail?
    d) How many children were asked **altogether**?

Understanding graphs and charts

2) Noah rolls a dice 30 times and records the following scores:

3 6 4 4 5 1
3 6 3 6 2 1
1 6 2 6 3 5
3 1 2 1 3 6
4 3 5 2 5 5

Score	Tally	Frequency
1		
2		
3		
4		
5		
6		

a) Copy and complete the frequency table.
b) How many times does Noah roll a 6?
c) Which score occurs most often?
d) How many **more** times does he score 5 than 4?

3) The bar chart shows the number of patients who visit a dentist.

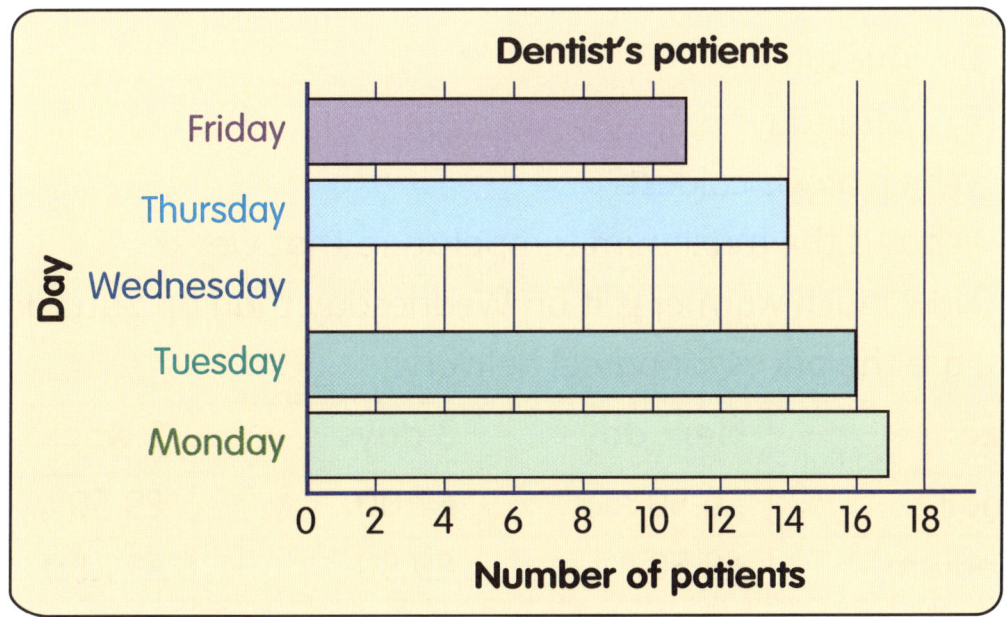

a) How many patients does the dentist see on Monday?
b) How many **more** patients does she see on Thursday than Friday?
c) How many patients does she see **in total**?
d) The dentist has a day off. Which day?

4) The line graph shows the temperature one week in April.

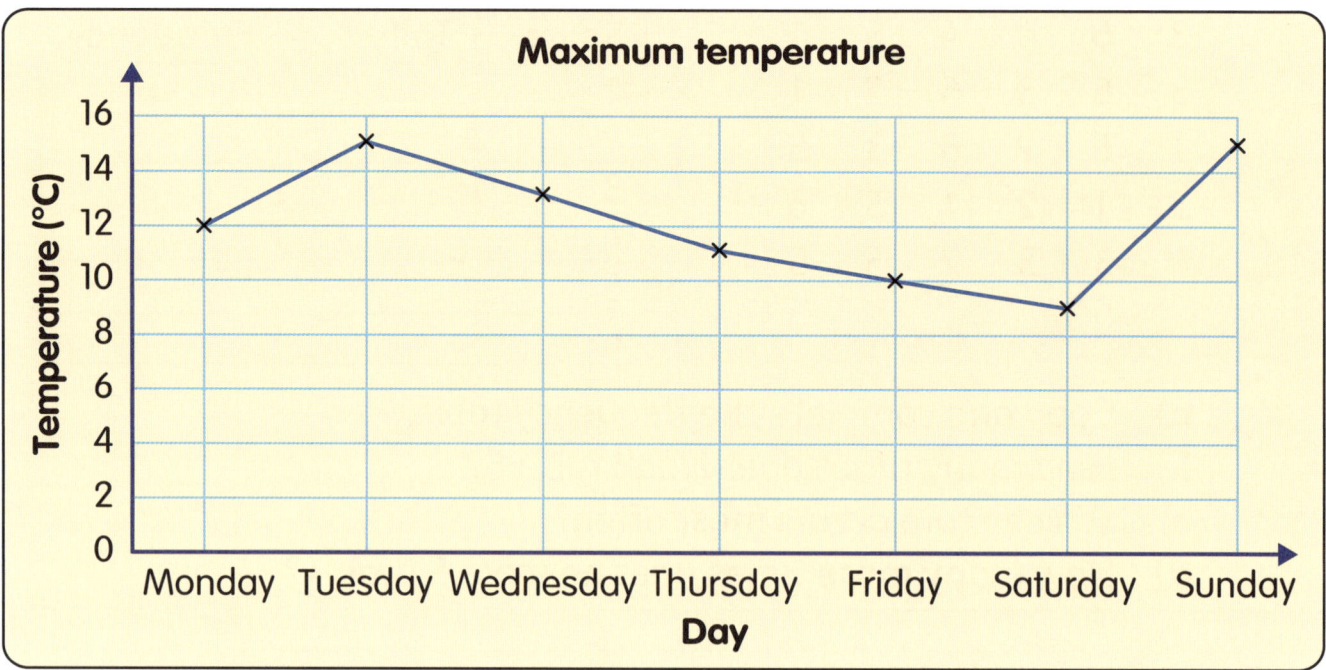

a) What is the temperature on:
   i) Tuesday
   ii) Saturday?
b) Which day is coldest?
c) What is the maximum temperature that week?
d) How much warmer is it on Wednesday than on Saturday?

5) Here are the prices for parcel delivery.

Size	Next day	3 days	1 week
small	£6.50	£7.00	£9.50
medium	£8.50	£9.50	£11.50
large	£9.50	£10.00	£13.50

a) How much does it cost to deliver:
  i) a small parcel the next day
  ii) a large parcel in 1 week
  iii) a medium parcel in 3 days?
b) Abdul pays exactly £10 for a parcel delivery.
   What is the size and how long will it take?

6) The comparative line graph shows the sales from two different car showrooms, Alfie Watts and Reg Motors.

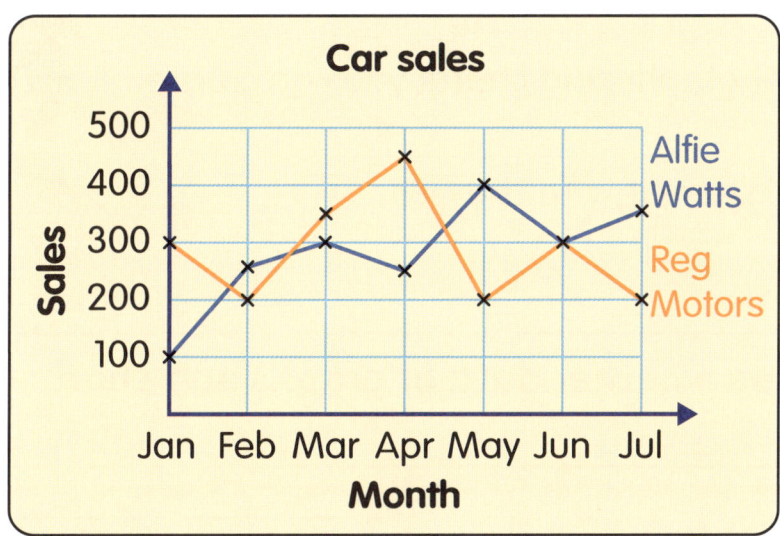

a) Which company has the better sales in:
  i) February   ii) April   iii) May   iv) June?
b) How many cars are sold by each company in:
  i) May   ii) March   iii) April?
c) Which company has the better (total) sales from January to July?

Chapter 16 Statistics

Now try this!

You can use technology to draw graphs and charts for data.

Spreadsheets often give you options to draw graphs and charts.

Enter the information about meals ordered in a restaurant into a spreadsheet.

	Adults	Children
pizza	15	19
pasta	18	6
salad	12	3

Investigate the different graphs and charts you can create using the spreadsheet.

Print your favourite and share it with the class.

Misleading data

 I will learn to decide whether data or graphs and charts are misleading.

The way information is given to you can be **misleading**. This might be because only a small number of people or things are recorded or because the graphs or charts drawn are not accurate.

Sometimes this is **deliberate**.

This bar chart shows the average price of a litre of petrol in 2020 and 2023.

It looks as if the price of petrol in 2023 was 3 times the price in 2020. In fact it is only 40p more.

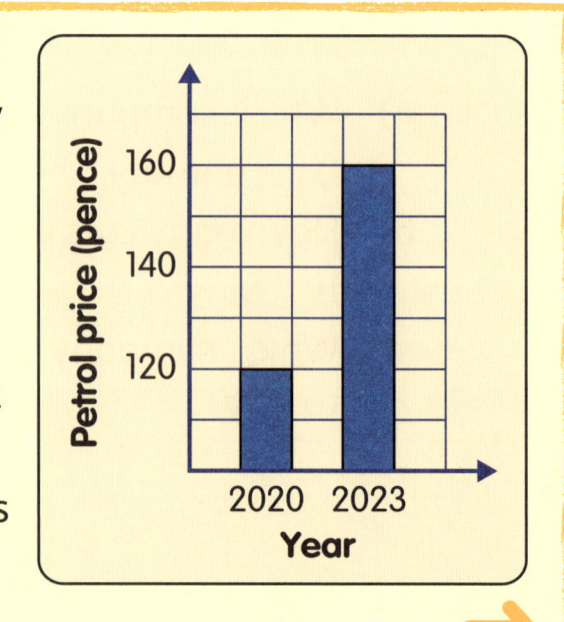

Understanding graphs and charts

The bar chart is misleading because the scale does not start at 0.

This bar chart shows a better representation.

There are many ways in which a graph or chart can be misleading. Look out for:
- no labelling on axes
- inaccurate labelling on axes
- sizes of boxes or symbols.

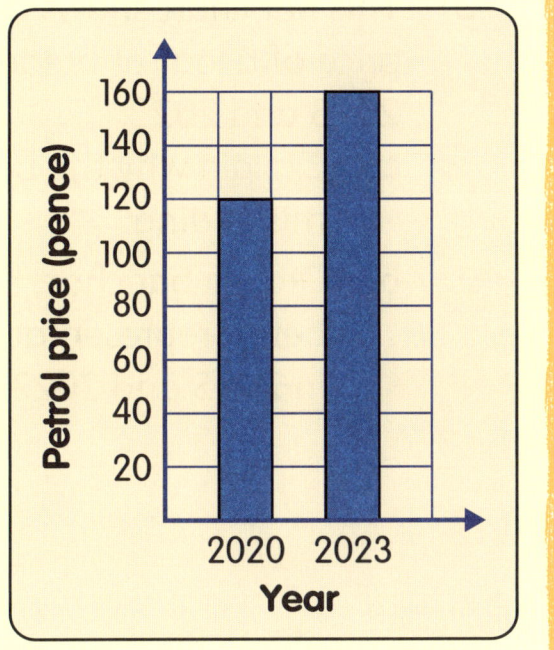

Exercise 2

1) Discuss with your teacher and the rest of the class the two bar charts shown in the example above.
    a) Why do you think the first bar chart is drawn in this misleading way?
    b) Who do you think might have wanted to show the price differences in this misleading way?

2) Hari asks 10 people in his class what their favourite sport is.
    Seven people say football.
    Hari says: 'The most popular sport in Scotland is football.'
    Do you think Hari has enough evidence to say this?
    Talk to a partner.

Chapter 16 Statistics

3) This bar chart shows the average price of a loaf of bread in a shop in 2015 and 2023.
   a) Explain why the bar chart is misleading.
   b) Draw a bar chart that gives a better representation of prices in 2015 and 2023.

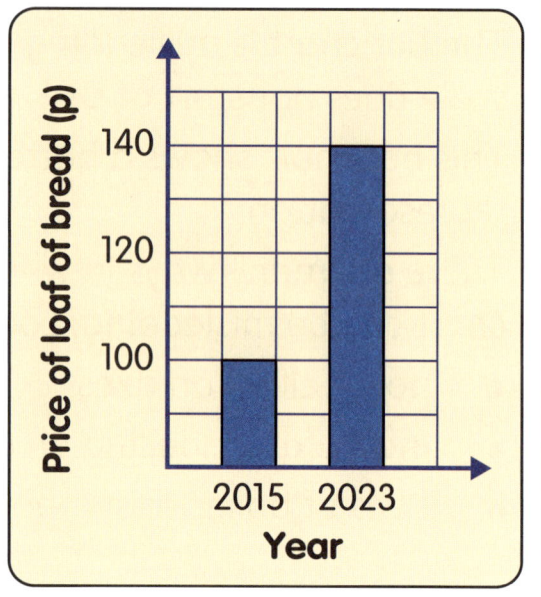

4) An IT company produces a line graph showing how a customer saves time using their new software.

   The customer used to spend 61.5 minutes a week generating reports. Over 10 weeks this went down by 2.5 minutes.

   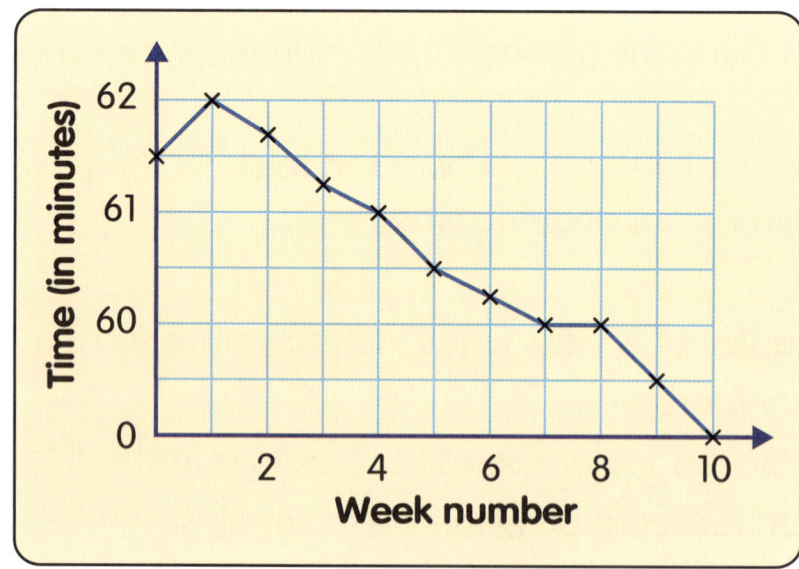

   a) Explain why the graph is misleading.
   b) Draw an accurate line graph.

5) Look at the following graphs and charts and decide what is wrong with them (there may be more than one point).

Explain how the graph or chart could be improved.

Look for:
- inaccurate labelling
- no labelling
- size of symbols
- uneven scales.

a)

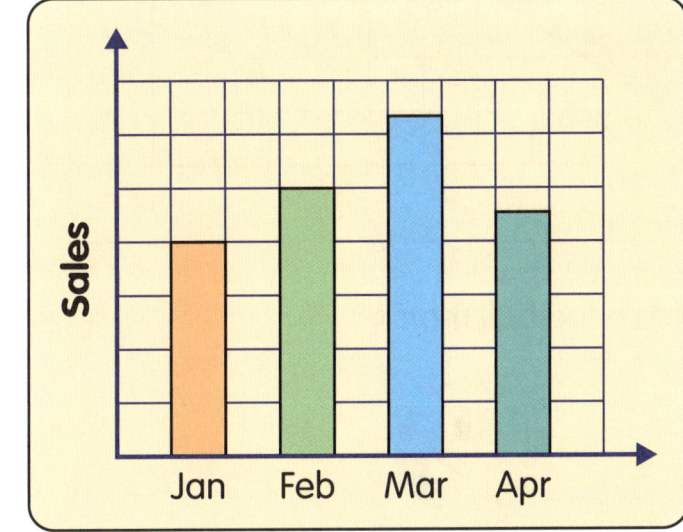

b)

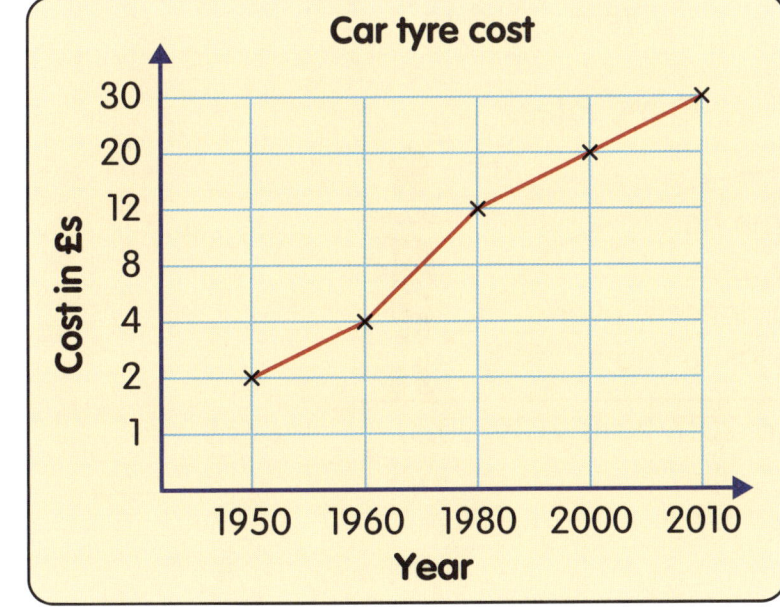

Chapter 16 Statistics

c)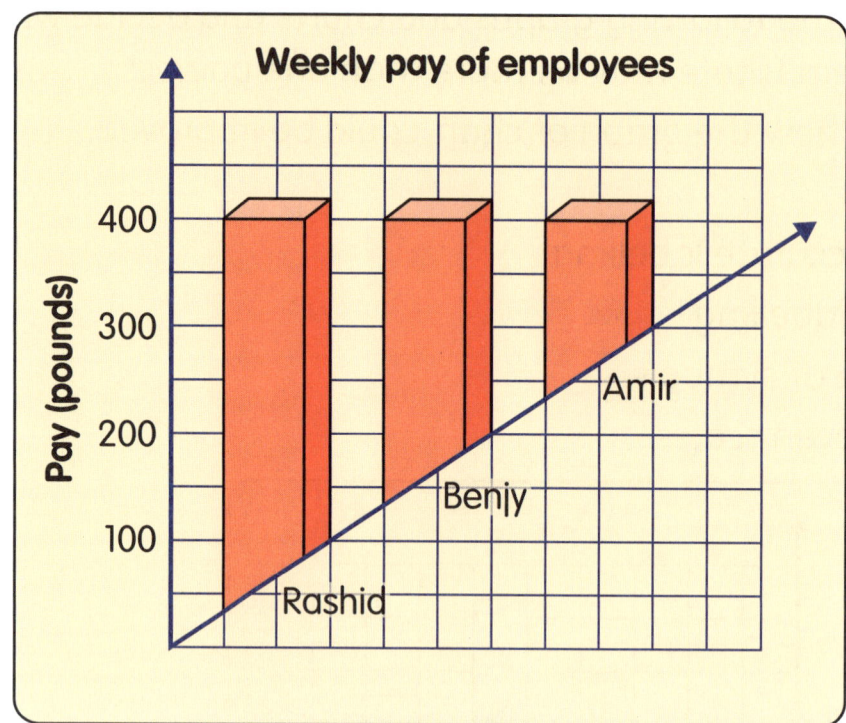

d)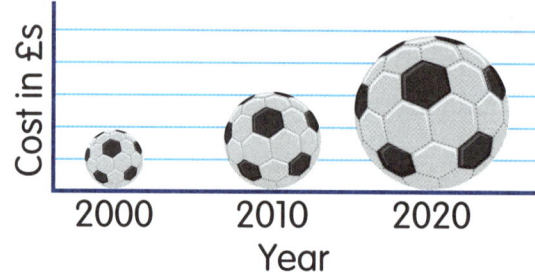

e)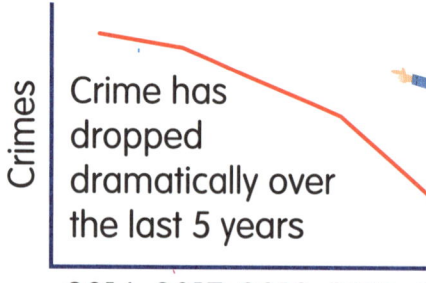

Understanding graphs and charts

Revisit, review, revise

1) The heartbeats of two tennis fans are monitored every 12 minutes during a match.

   The results are shown in the line graph:

   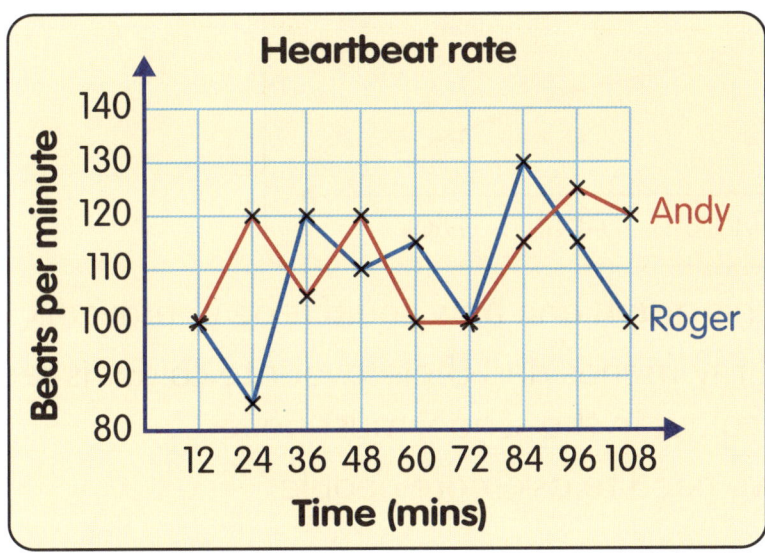

   a) How many times is Roger's heartbeat:
      i) above Andy's
      ii) the same as Andy's
      iii) below Andy's?
   b) In the 48th minute, what is the difference between Roger's and Andy's heartbeats?

2) The table shows the number of washing machines sold by two shops:

Month	Jan	Feb	Mar	Apr	May	Jun
Alectric	11	13	16	20	17	13
Scotlec	6	6	13	24	22	15

   a) How many washing machines did Scotlec sell in April?
   b) How many **more** washing machines did Alectric sell than Scotlec in February?
   c) Which company sold more washing machines in **total**?

Chapter 16 Statistics

3) The line graph shows the number of cases of an infection in a country.

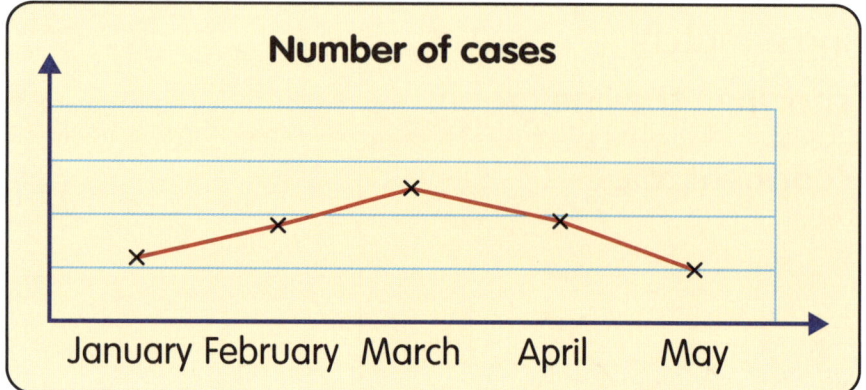

What is wrong with this graph and how could it be improved?

4) Ami says: 'I asked all my friends and their favourite colour is red, so the most popular colour in the school must be red.'

Darsh says: 'I think we need to ask more people.'

Who is correct?

Explain why.

17 Whole numbers 2
Big numbers, rounding and estimating

Whole numbers to 1 000 000

 I will learn to understand place value for numbers up to 1 000 000

Remember, remember
The number after 9999 is called **ten thousand**.
It is written as **10 000**
10 thousands = 10 000

After 9999 is 10 000 (ten thousand),
then 10 001, 10 002, 10 003 ... 10 999

After 10 999 is 11 000 (eleven thousand),
then 11 001, 11 002, 11 003 ... 11 999

After 11 999 is 12 000 (twelve thousand),
then 12 001, 12 002, 12 003 ... 12 999

This pattern goes on through 13 000, 14 000, 15 000 and so on up to 99 999.

The number after 99 999 (ninety-nine thousand, nine hundred and ninety-nine) is called **one hundred thousand**.
It is written as **100 000**
100 thousands = 100 000

After 99 999 is 100 000 (one hundred thousand),
then 100 001, 100 002, 100 003 ... 100 999

After 100 999 is 101 000 (one hundred and one thousand),
then 101 001, 101 002, 101 003 ... 101 999

Chapter 17 Whole numbers 2

After 101 999 is 102 000 (one hundred and two thousand), then 102 001, 102 002, 102 003 … 102 999

This pattern goes on through 103 000, 104 000, 105 000 and so on up to 999 999

The number after 999 999 (nine hundred and ninety-nine thousand, nine hundred and ninety-nine) is called **one million**.

It is written as **1 000 000**

1000 thousands = 1 000 000

Example

What does each digit stand for in the number **346 785**?

- the **3** stands for three hundred thousand   300 000
- the **4** stands for forty thousand   40 000
- the **6** stands for six thousand   6000
- the **7** stands for seven hundred   700
- the **8** stands for eight tens   80
- the **5** stands for five ones   5

The number 346 785 in words is:

three hundred and forty-six thousand, seven hundred and eighty-five.

Exercise 1

 1) In your jotter, copy and complete:

a) 15 824 = 10 000 + 5000 + ____ + 20 + ____

b) 26 193 = 20 000 + ____ + 100 + ____ + ____

c) 74 306 = ____ + ____ + ____ + 6

d) 631 482 = 600 000 + 30 000 + ____ + 400 + ____ + ____

e) 498 103 = ____ + 90 000 + ____ + ____ + ____

Big numbers, rounding and estimating

2) What do these digits stand for in the number **26 915**?
   a) 2   b) 6   c) 9   d) 1   e) 5

3) What do these digits stand for in the number **923 461**?
   a) 9   b) 2   c) 3   d) 4   e) 6   f) 1

4) What does the **7** stand for in each number?
   a) 58 740   b) 35 279   c) 647 900   d) 875 348

5) Copy and complete with words:
   a) 50 000 = _____ thousand
   b) 16 742 = sixteen thousand, seven _____ and _____
   c) 81 620 = eighty-one _____, six _____ and _____
   d) 27 105 = _____ thousand, _____ hundred and _____
   e) 300 000 = three _____ thousand
   f) 113 259 = one hundred and _____ thousand, _____ hundred and _____
   g) 1 000 000 = one _____

6) In your jotter, write these numbers in **words**:
   a) 21 900   b) 637 350   c) 580 400   d) 800 101

7) In your jotter, write these numbers in **digits**:
   a) forty-three thousand, two hundred and ninety
   b) two hundred and thirty thousand, four hundred and sixty-one
   c) one million
   ☀ d) six hundred thousand and four.

Chapter 17 Whole numbers 2

8) In your jotter, write the **bigger** number in each pair.
   a) 104 374 and 257 222
   b) 56 731 and 55 841
   c) 98 411 and 98 114
   d) 109 999 and 1 000 000

9) In your jotter, write each group of numbers in order of size. Start with the **smallest**.
   a) 71 682   6876   70 186   663 809
   b) 100 870   99 924   100 086   98 999   90 887
   c) 308 000   83 000   838 000   38 888   883 003

 10) In 2021, the population of Dundee City was 147 720.
In 2020 there were 1100 more people.
What was the population of Dundee City in 2020?

Rounding large numbers

 I will learn to round tens of thousands and hundreds of thousands to the nearest 1000, 10 000 and 100 000.

Remember, remember

The **last thousand** before 3287 is 3000.
The **next thousand** after 3287 is 4000.
3287 is closer to 3000 than to 4000.

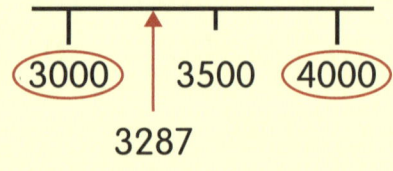

We say that '3287, **rounded to the nearest 1000**, is 3000' or '3287 **rounds down** to 3000, to the nearest 1000'.
Notice that 3500 is the same distance from 3000 and 4000.
We say that '3500 **rounds up** to 4000 to the nearest 1000'.

Big numbers, rounding and estimating

Rounding **tens of thousands** and **hundreds of thousands** to the nearest 10 000 and 100 000 follows the same rules as rounding to the nearest 1000.

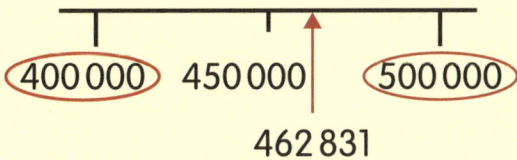

462 831

The **last hundred thousand** before 462 831 is 400 000.
The **next hundred thousand** after 462 831 is 500 000.

462 831 is closer to 500 000 than to 400 000.

We say that '462 831, **rounded to the nearest 100 000**, is 500 000' or '462 831 **rounds up** to 500 000, to the nearest 100 000'.

450 000 **rounds up** to 500 000 to the nearest 100 000 too.

Another way to round to the **nearest 100 000** is to look at the **ten thousands digit**:

4(6)2 831   4(5)0 000

- If the ten thousands digit is **5, 6, 7, 8 or 9**, round up.
- If the ten thousands digit is **0, 1, 2, 3 or 4**, round down.

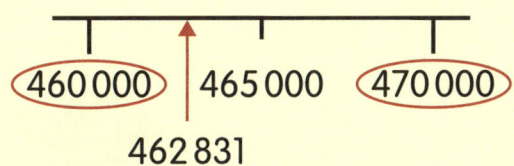

462 831

The **last ten thousand** before 462 831 is 460 000.
The **next ten thousand** after 462 831 is 470 000.

462 831 is closer to 460 000 than to 470 000.

We say that '462 831, **rounded to the nearest 10 000**, is 460 000' or '462 831 **rounds down** to 460 000 to the nearest 10 000'.

465 000 **rounds up** to 470 000 to the nearest 10 000.

Chapter 17 Whole numbers 2

Another way to round to the **nearest 10 000** is to **look at the thousands** digit:

46②831   46⑤000

- If the thousands digit is **5, 6, 7, 8** or **9**, round up.
- If the thousands digit is **0, 1, 2, 3** or **4**, round down.

Exercise 2

1) In your jotter, write the **missing** numbers.
   a) The last ten thousand before 11 573 is 10 000.
      The next ten thousand after 11 573 is ___.
   b) The last ten thousand before 38 710 is ___.
      The next ten thousand after 38 710 is ___.
   c) The last hundred thousand before 291 453 is 200 000.
      The next hundred thousand after 291 453 is ___.
   d) The last hundred thousand before 628 910 is ___.
      The next hundred thousand after 628 910 is ___.
   e) The last ten thousand before 628 910 is 620 000.
      The next ten thousand after 628 910 is ___.

2) In your jotter, write the **missing** numbers.
   Use the number lines to help you.
   a) 34 120 rounds to ___ to the nearest 10 000.

   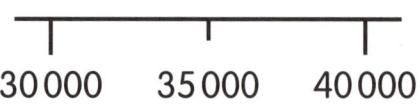

   b) 782 405 rounds to ___ to the nearest 100 000.

   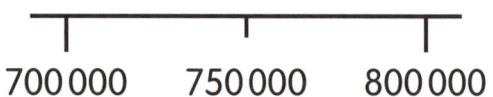

Big numbers, rounding and estimating

c) 782 405 rounds to ____ to the nearest 10 000.

780 000	785 000	790 000

3) a) Copy this number line.

400 000	450 000	500 000

b) Mark 410 567 on your number line.
c) Does 410 567 **round up** or **round down** to the nearest 100 000?
d) What is 410 567 rounded to the nearest 100 000?

4) a) Copy the number 410 567.
b) Circle the **ten thousands** digit.
c) Does your circled number tell you to **round up** or **round down** to the **nearest 100 000**?
d) Does your answer to part c) agree with your answer to question 3c)? Yes or no.

5) a) Copy this number line.

210 000	215 000	220 000

b) Mark 217 402 on your number line.
c) Does 217 402 **round up** or **round down** to the nearest 10 000?
d) What is 217 402 rounded to the nearest 10 000?

6) a) Copy the number 217 402.
b) Circle the **thousands** digit.
c) Does your circled number tell you to **round up** or **round down** to the **nearest 10 000**?
d) Does your answer to part c) agree with your answer to question 5c)? Yes or no.

Chapter 17 Whole numbers 2

7) Round these numbers to the nearest 100 000:
   a) 670 000
   b) 532 000
   c) 194 800
   d) 491 003
   e) 350 000
    f) 971 237

8) Round these numbers to the nearest 10 000:
   a) 751 000
   b) 386 009
   c) 638 710
   d) 920 100
   e) 170 992
    f) 996 341

9) 700 012 people used Inverness airport in 2022. Round this number of people to the nearest:
   a) 100 000
    b) 10 000

Estimating by rounding

 I will learn to estimate answers by rounding.

Remember, remember

You have already learnt to round numbers to the nearest 100 or 1000 to estimate an answer.

Example 1
620 + 174
is about 600 + 200 = 800

Example 2
7431 – 4910
is about 7000 – 5000 = 2000

Big numbers, rounding and estimating

You can round numbers in a calculation to the nearest 10 000 or 100 000 to estimate the answer.

Example 1

25 417 + 21 925

is about:

30 000 + 20 000 = 50 000

Example 2

920 414 − 470 641

is about:

900 000 − 500 000 = 400 000

The calculations for an estimated answer are easier when **only one digit is not zero**.

Exercise 3

1) In your jotter, write:
   a) 28 765 to the nearest 10 000
   b) 52 466 to the nearest 10 000
   c) Use your answers to a) and b) to copy and complete:
      28 765 + 52 466 is about ____ + ____ = ____

2) In your jotter, write:
   a) 742 321 to the nearest 100 000
   b) 602 599 to the nearest 100 000
   c) Use your answers to a) and b) to copy and complete:
      742 321 − 602 599 is about ____ − ____ = ____

3) Copy and complete to estimate the answer to each calculation:
   a) 91 064 − 18 471 is about ____ − 20 000 = ____
   b) 473 471 + 339 662 is about 500 000 + ____ = ____
   c) 35 471 + 48 006 is about ____ + ____ = ____
   d) 897 144 − 103 765 is about ____ + ____ = ____

Chapter 17 Whole numbers 2

4) In your jotter, estimate the answer to each calculation.

a) 29 510 − 11 047

b) 81 476 − 19 600

 c) 676 588 − 42 418

 d) 32 997 + 9961

5) In one year, 247 200 cars used the ferry between Ardgour and Nether Lochaber.

In the same year, 599 600 cars used the ferry between Gourock and Dunoon.

Estimate how many **more** cars used the Gourock/Dunoon ferry.

 6) A house is for sale for £448 500.

The house next door is for sale for £457 500.

Estimate whether the **total** cost of buying both houses is less than £1 000 000.

Show your working.

Now try this!

Work with a partner.

It is estimated that in Scotland there are:
- 8000 otters
- 400 000 red deer
- 120 000 red squirrels.

For each fact, do you think the number is rounded to the nearest 1000, 10 000 or 100 000?

Write two possible numbers for the population of each animal.

Big numbers, rounding and estimating

Revisit, review, revise

1) In your jotter, copy and complete:
   508 631 = ___ + ___ + 600 + ___ + ___.

2) What do these digits stand for in the number **178 061**?
   a) first 1   b) 7   c) 8   d) 6   e) last 1

3) What does the **9** stand for in each number?
   a) 10 903   b) 29 406   c) 917 100

4) In your jotter, write these numbers in words:
   a) 32 608   b) 570 302

5) In your jotter, write **forty-three thousand and seventy** in digits.

6) In your jotter, write these numbers in order of size.
   Start with the **smallest**.

   | 84 770 | 87 000 | 84 777 | 78 888 | 86 999 |

7) Copy and complete:
   a) the last hundred thousand before 307 885 is ___
   b) the next hundred thousand after 307 885 is ___
   c) 307 885 rounded to the nearest 100 000 is ___

8) Round these numbers to the nearest 10 000:
   a) 73 410   b) 29 004   c) 687 314

9) In your jotter, estimate the answer to each calculation.
   a) 35 671 + 47 471   b) 751 481 − 275 714

10) One morning, 42 641 people use Glasgow Central Station.
    That afternoon and evening, 45 164 people use the station.
    Estimate how many people use the station that day.

11) A company makes a profit of £102 567 in its first year.
    The same company makes a profit of £197 417 in its second year.
    Estimate how much **more** profit it makes in its second year than in its first year.

18 Symmetry
Identifying and completing symmetrical shapes and patterns

Symmetrical shapes and patterns

> I will learn to identify lines of symmetry and complete symmetrical patterns.

Remember, remember

A shape has a **line of symmetry** if, when you fold the shape over the line, the two halves exactly match.

A shape can have **0** lines of symmetry:

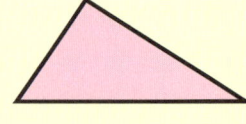

A shape can have **1** line of symmetry:

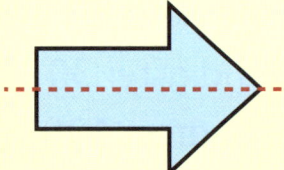

A shape can have **more than 1** line of symmetry:

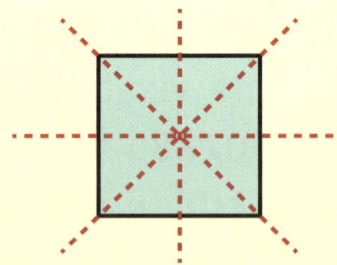

If you are given half of a symmetrical shape on a grid with a line of symmetry, draw the other half by reflecting it in the line of symmetry.

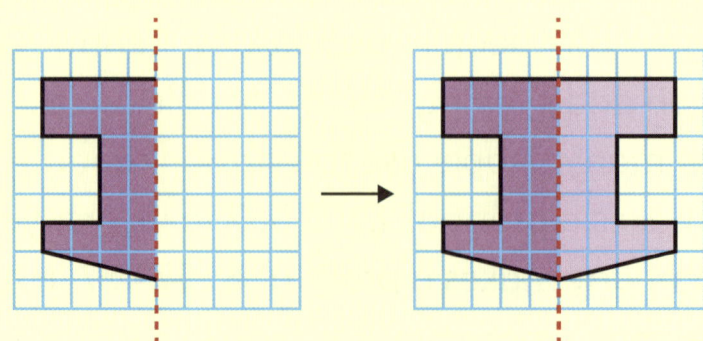

Identifying and completing symmetrical shapes and patterns

To complete symmetrical patterns, reflect colours and shapes in the lines of symmetry.

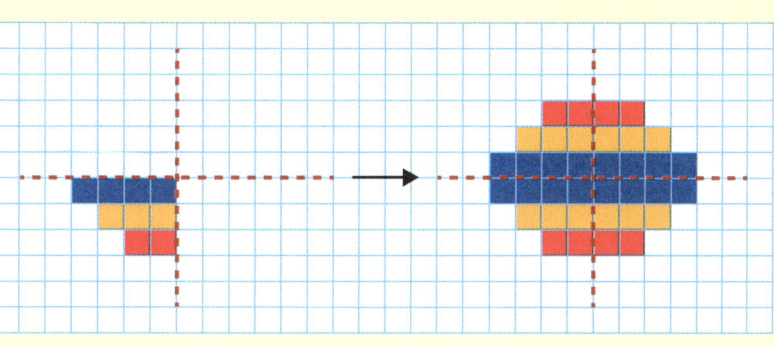

Exercise 1

1) Copy or trace these shapes and draw on all the lines of symmetry:

   a) b) c) d) e) f)

2) How many lines of symmetry does each image have?

   a) 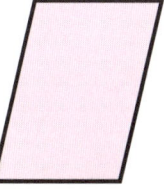　　b) 　　c)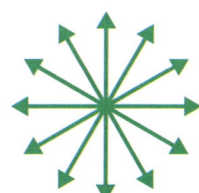

   d) 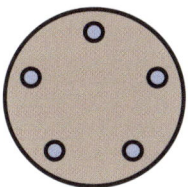　　e) 　　f)

Chapter 18 Symmetry

3) Copy these shapes onto squared paper.
   Complete the shapes so that the dashed lines are lines of symmetry.

   a)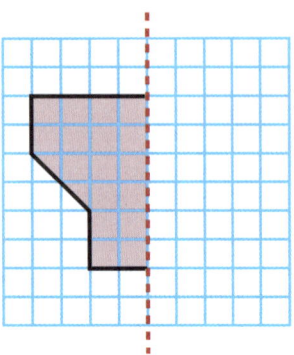

   b)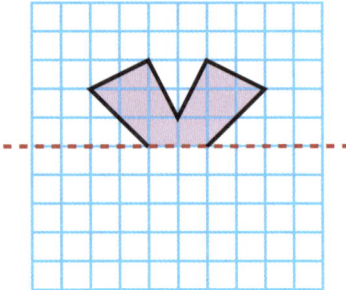

   c)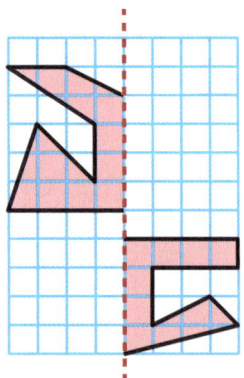

   d)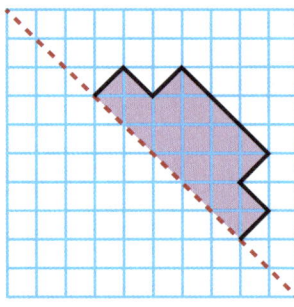

4) Copy and complete the symmetrical patterns.

   a) b) c) d) e) f)

   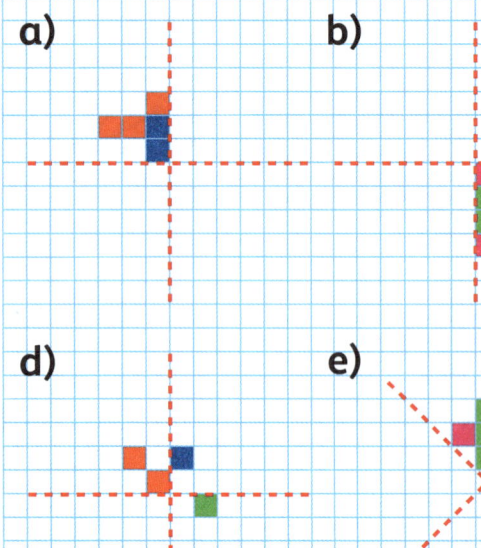
   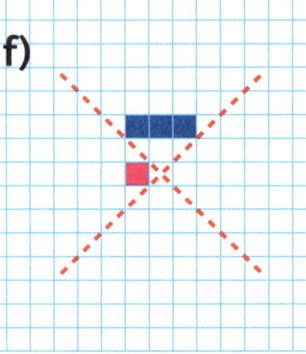

Identifying and completing symmetrical shapes and patterns

5) How many lines of symmetry do these shapes have?
   (It might help to draw them.)
   a) an isosceles triangle
   b) an equilateral triangle
   c) a scalene triangle
   d) a rectangle
   e) a square

6) A regular pentagon has 5 lines of symmetry.

   How many lines of symmetry does a regular hexagon have?

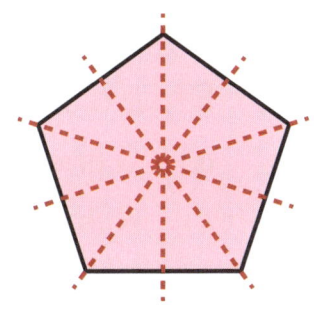

Now try this!

Symmetrical patterns are everywhere in nature.
Write a list of places you may (or may not) see symmetry in animals, plants and flowers.

Chapter 18 Symmetry

Revisit, review, revise

1) Copy or trace these shapes and draw in any lines of symmetry:

   a) 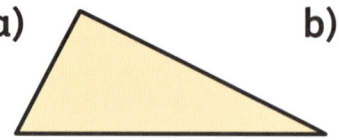   b)    c)    d)

2) Draw a shape with:

   a) 0 lines of symmetry

   b) 1 line of symmetry

   c) 4 lines of symmetry.

3) Copy and complete each shape so that the dashed line is a line of symmetry.

   a) 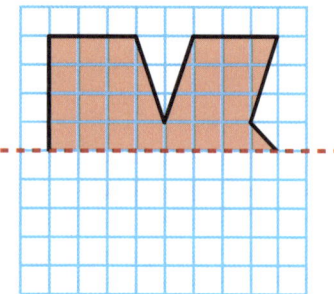   b)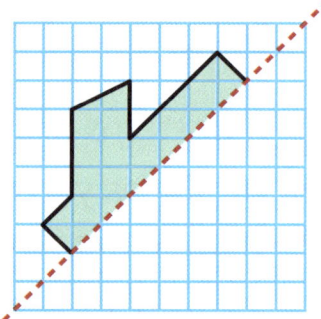

4) Copy and complete the symmetrical patterns.

   a) 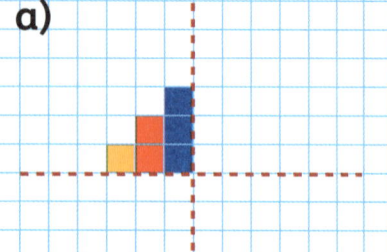   b) 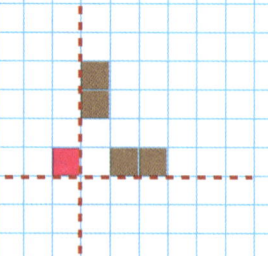   c)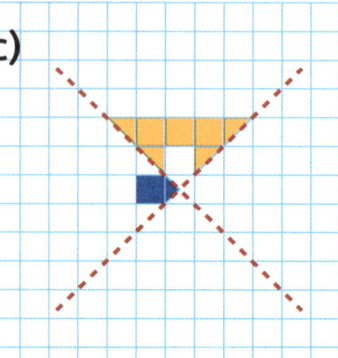

19 Sequences 2
More sequences

Rules for sequences

> I will learn to identify a linear sequence and find a rule for extending it.

Remember, remember

A **sequence** is a list of numbers or objects in a given order.

Each number or object in a sequence is called a **term**.

One way to describe a sequence is to give the **first term** and the **term-to-term rule**.

You can describe the sequence

4, 8, 12, 16, …

by writing: first term = 4; term-to-term rule = +4.

Finding a general rule (or **formula**) for a sequence makes it easy to find any term in the sequence.

It can be helpful to draw a **table** to spot a rule for a sequence.

Here is a sequence showing cakes.

Each cake has 4 candles.

Draw a table to show the number of cakes and number of candles:

Number of cakes	1	2	3	4	5
Number of candles	4	8	12	__	__

Chapter 19 Sequences 2

> The number of candles is 4 times the number of cakes.
> To find any term in the sequence, use the formula:
> **number of candles = 4 × number of cakes**
> For **5** cakes you would need:
> number of candles = 4 × **5** = 20 candles.

Exercise 1

1) In a library, 3 people can sit around each table.

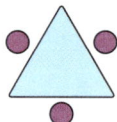   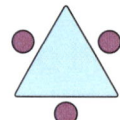   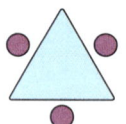 

   1 table   2 tables   3 tables
  3 people   6 people   9 people

   a) Copy and complete the table.

Number of tables	1	2	3	4	5
Number of people	3	6	___	___	___

   b) For every extra table, how many extra people are seated?
   c) Copy and complete the formula:
      number of people = ____ × number of tables.
   d) How many people can sit at 20 tables?

2) Here are the prices of footballs.

    1 ball   2 balls   3 balls
     £7   £14   £21

More sequences

a) Copy and complete the table.

Number of footballs	1	2	3	4	5
Price	__	__	__	__	__

b) Copy and complete the formula:
cost = ___ × number of footballs.

c) A football club buys 30 footballs.
How much do they cost?

d) Another club spends £84 on footballs.
How many does it buy?

3) A tiling pattern is made using blue and red tiles.

Pattern 1  Pattern 2  Pattern 3

a) Copy and complete the table.

Pattern number	1	2	3	4	5
Number of blue tiles	1	2	3	__	__
Number of red tiles	2	4	__	__	__

b) Copy and complete:
number of blue tiles = ___ × pattern number
number of red tiles = ___ × pattern number.

c) In the 20th pattern, how many:
   i) blue tiles   ii) red tiles   iii) tiles?

d) How many tiles altogether in the 25th pattern?

Chapter 19 Sequences 2

4) Here is a recipe for 10 gingerbread people:

butter..............................100 g
sugar..............................100 g
flour...............................250 g
icing sugar......................50 g
ginger..............................5 g

a) How much butter is needed for 1 gingerbread person?

b) How much flour is needed for 1 gingerbread person?

c) Copy and complete the table.

Number of gingerbread people	1	2	3	4	5
Butter	___	___	___	___	___
Sugar	10 g	20 g	___	___	___
Flour	___	___	___	___	___
Icing sugar	5 g	___	___	___	___
Ginger	0.5 g	___	___	___	___

d) Copy and complete:

butter = ___ × number of gingerbread people

sugar = ___ × number of gingerbread people

flour = ___ × number of gingerbread people

icing sugar = ___ × number of gingerbread people

ginger = ___ × number of gingerbread people.

e) Ollie makes 25 gingerbread people.
How much of each ingredient does he need?

f) Fara uses 200 g of flour to make gingerbread people.
How many does she make?

More sequences

5) For each of the tables below, complete the formula connecting the two letters and copy and complete the table.

a) P = _____ × N

No. of newspapers (N)	1	2	3	4	5	6
No. of pages (P)	30	60	90	___	___	___

b) P = _____ × T

No. of trees (T)	1	2	3	4	5	6
No. of pineapples (P)	18	36	54	___	___	___

c) H = _____ × D

No. of days (D)	1	2	3	4	5	6
No. of hours (H)	24	48	72	___	___	___

d) C = _____ × _____

No. of muffins (M)	2	3	4	5	6	7
Cost in £s (C)	2.50	3.75	5.00	___	___	___

e) _____ = _____ × _____

No. of jars (J)	3	4	5	6	7	8
No. of jelly beans (B)	450	600	750	___	___	___

Chapter 19 Sequences 2

Square and triangular numbers

> 💡 I will learn what square and triangular numbers are.

The number 9 is called a **square number**, because 9 dots can be arranged to make a square pattern (3 by 3).

The sequence of square numbers is: 1, 4, 9, 16, 25 …

1   4   9   16   25

Exercise 2

1) a) Draw the next square dot pattern after 25.
   b) What is the next square number after 25?
   c) What is the 7th square number?
   d) What is the 10th square number?

More sequences

2) Draw or print a hundred grid and shade the square numbers.

1	2	3	4	5	6	7	8	9	10
11	12	13	14	15	16	17	18	19	20
21	22	23	24	25	26	27	28	29	30
31	32	33	34	35	36	37	38	39	40
41	42	43	44	45	46	47	48	49	50
51	52	53	54	55	56	57	58	59	60
61	62	63	64	65	66	67	68	69	70
71	72	73	74	75	76	77	78	79	80
81	82	83	84	85	86	87	88	89	90
91	92	93	94	95	96	97	98	99	100

3) Look at the sequence of square numbers.

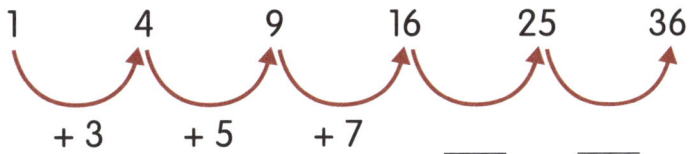

1   4   9   16   25   36

+ 3   + 5   + 7   ___   ___

a) What do you notice?

b) Continue the sequence up to 100.

Triangular patterns of dots give the **triangular numbers**.
The sequence of triangular numbers is: 1, 3, 6, 10 …

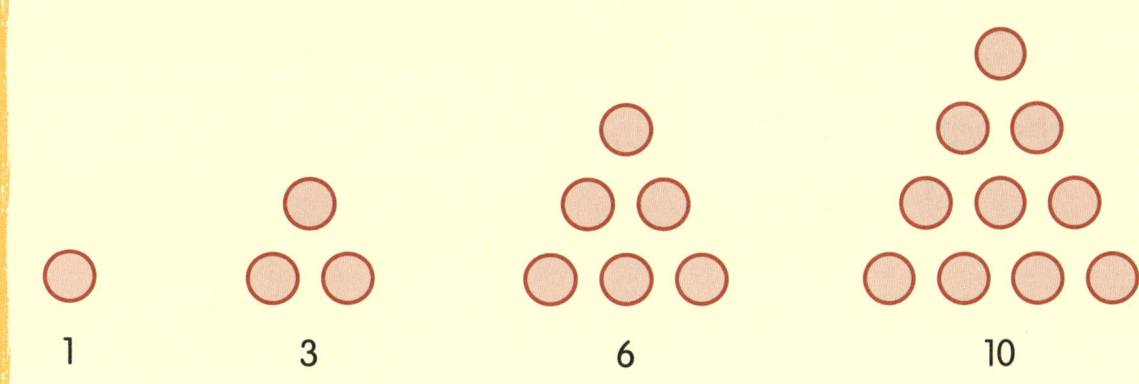

1   3   6   10

Chapter 19 Sequences 2

4)
- **a)** Draw the next triangular dot pattern after 10.
- **b)** What is the next triangular number after 10?
- **c)** What is the 6th triangular number?
- **d)** What is the 7th triangular number?

5) Draw or print a hundred grid and shade the triangular numbers.

1	2	3	4	5	6	7	8	9	10
11	12	13	14	15	16	17	18	19	20
21	22	23	24	25	26	27	28	29	30
31	32	33	34	35	36	37	38	39	40
41	42	43	44	45	46	47	48	49	50
51	52	53	54	55	56	57	58	59	60
61	62	63	64	65	66	67	68	69	70
71	72	73	74	75	76	77	78	79	80
81	82	83	84	85	86	87	88	89	90
91	92	93	94	95	96	97	98	99	100

6) Look at the sequence of triangular numbers.

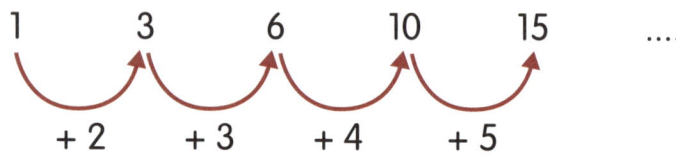

1   3   6   10   15   ....

+ 2   + 3   + 4   + 5

- **a)** What do you notice?
- **b)** Continue the sequence up to 100.

7) Look at the pattern:

1 = 1
1 + 2 = 3
1 + 2 + 3 = 6
1 + 2 + 3 + 4 = 10

Continue the pattern for 5 more rows.

More sequences

Now try this!

Can you make rectangular numbers?

Sam makes a sequence of rectangles.

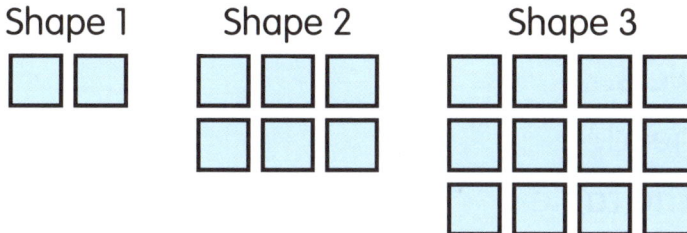

Shape 1 Shape 2 Shape 3

1) How many squares will he need to make the next rectangle?

2) Can you find a formula for the number of squares he uses in the sequence of shapes?

3) Design your own different sequence of rectangles and find a formula for the number of squares needed for terms in your sequence.

Revisit, review, revise

1) Each vase contains 5 flowers.

a) Copy and complete the table.

Number of vases	1	2	3	4	5
Number of flowers	___	___	___	___	___

b) Copy and complete the formula:
number of flowers = ____ × number of vases

c) Anita has 20 vases.
How many flowers will she need?

d) Muhammad has 135 flowers.
How many vases will he need?

2) Jen's rate of pay is shown in the table.

Number of hours (H)	1	2	3	4	5
Wage in £s (W)	8.20	16.40	24.60	____	____

a) How much is she paid for:
   i) 4 hours
   ii) 5 hours?

b) Copy and complete the formula:
wage = ____ × number of hours

c) One day Jen is paid £90.20.
How many hours does she work?

3) In your jotter, write all the **square numbers** between 30 and 60.

4) Copy and complete the table showing the **triangular numbers**.

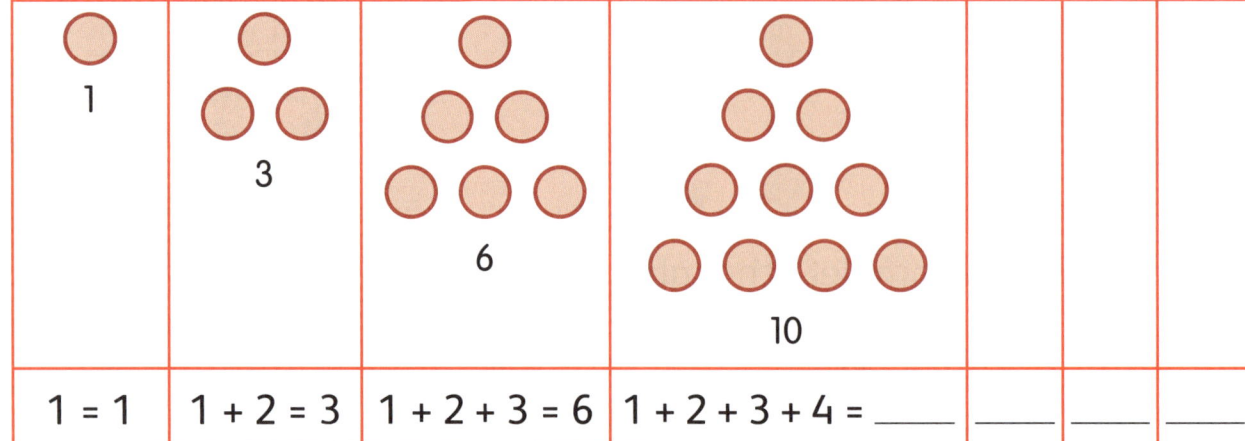

20 Coordinates
The coordinate axes

Plotting coordinates

 I will learn to plot points on the coordinate axes.

Remember, remember

The **coordinates** of M are (5, 7).
M is 5 along the *x*-axis and 7 up the *y*-axis.

The point (0, 0) is called the **origin**.

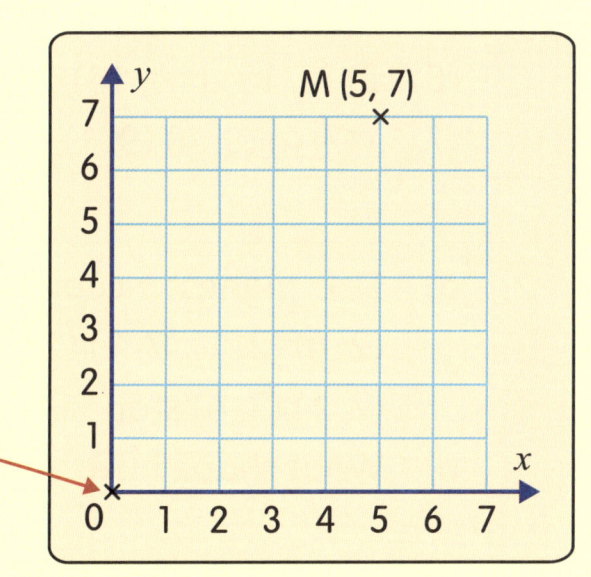

Pictures can be drawn using coordinates.
You can draw a right-angled triangle by plotting the points:

(1, 5)   (5, 5)   (5, 2)   (1, 5) stop.

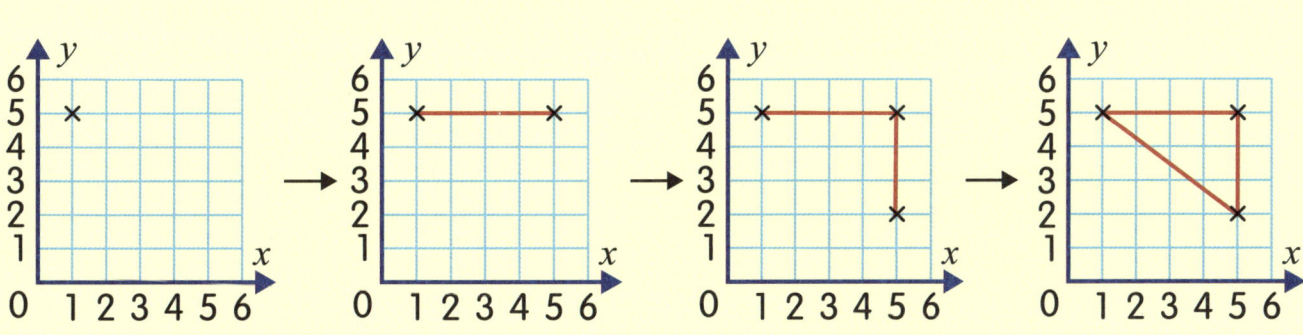

Chapter 20 Coordinates

Exercise 1

Draw a coordinate grid for each question. You are told what size.
Plot the points in order and join them as you go.

1) Size: highest x-coordinate 15; highest y-coordinate 15.
   (4, 1) (4, 11) (2, 11) (8, 15) (14, 11) (12, 11) (12, 1) (4, 1) stop
   (5, 7) (5, 9) (7, 9) (7, 7) (5, 7) stop
   (9, 7) (9, 9) (11, 9) (11, 7) (9, 7) stop
   (6, 12) (6, 13) (10, 13) (10, 12) (6, 12) stop
   (7, 1) (7, 4) (9, 4) (9, 1) stop

2) a) Size: highest x-coordinate 15; highest y-coordinate 10.
   (2, 4) (2, 1) (7, 1) (13, 4) (13, 7) (7, 4) (2, 4) (8, 7) (13, 7) stop
   (7, 1) (7, 4) stop
   b) What shape have you drawn?

3) a) Size: highest x-coordinate 15; highest y-coordinate 10.
   (2, 1) (2, 9) (15, 9) (15, 1) (2, 1) stop
   (3, 2) (3, 8) (14, 8) (14, 2) (3, 2) stop
   (4, 3) (4, 7) stop
   (5, 7) (5, 3) (7, 3) (7, 5) (5, 7) stop
   (8, 7) (10, 7) (10, 5) (8, 5) (8, 3) (10, 3) stop
   (11, 3) (11, 7) (13, 7) (13, 3) (11, 3) stop
   b) What time is it?

4) In your jotter, write the instructions for drawing the word MATHS.

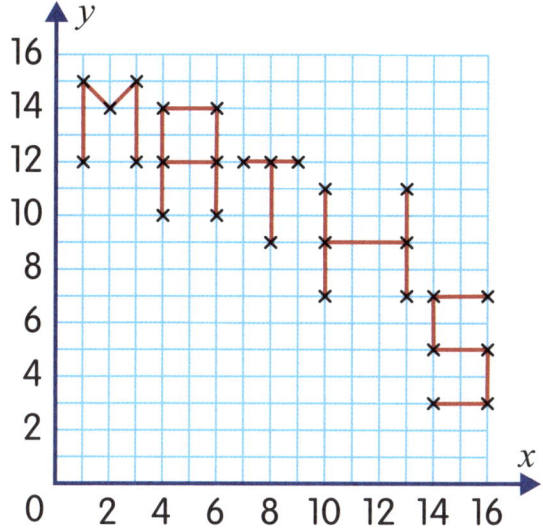

Now try this!

A coordinate challenge

Can you position these ten letters in the correct places, according to the eight clues below?

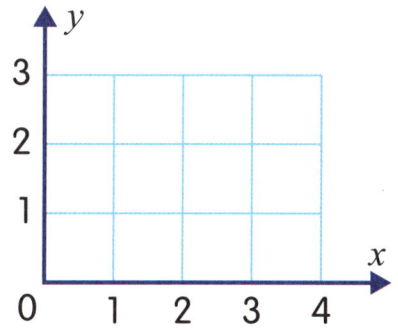

Clues:
- The letters at (1, 1), (1, 2) and (1, 3) have a vertical line of symmetry.
- The letters at (1, 1), (2, 1) and (3, 1) are symmetrical about a horizontal line.

Chapter 20 Coordinates

- The letter at (4, 2) is P.
- The letter at (3, 1) is made of only straight lines.
- The letters at (3, 3) and (2, 0) are made of only curved lines.
- The letters at (3, 3), (3, 2) and (3, 1) are consecutive in the alphabet.
- The letters at (0, 2) and (1, 2) are at the two ends of the alphabet.

Coordinates and shapes

 I will learn to plot shapes on the coordinate axis.

Remember, remember

Polygons are 2D shapes with straight sides.

Number of sides	3	4	5	6
Name of polygon	triangle	quadrilateral	pentagon	hexagon

A **right-angled triangle** has a 90° angle.

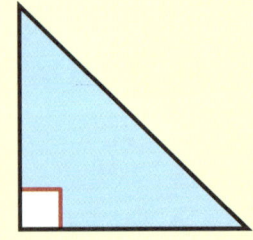

The coordinate axes

An **isosceles triangle** has two sides equal and two angles equal.

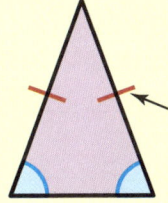
← these lines tell you which sides are the same length

An **equilateral triangle** has all sides equal and all angles equal.

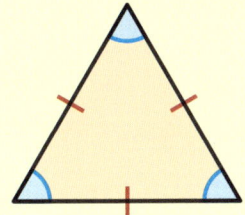

A **scalene triangle** has no sides equal and no angles equal.

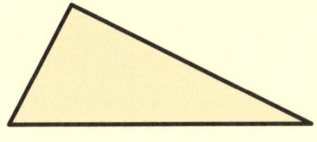

You can draw shapes on the **coordinate grid**.
Rectangle ABCD has coordinates:

A (2, 7)
B (11, 7)
C (11, 4)
D (2, 4)

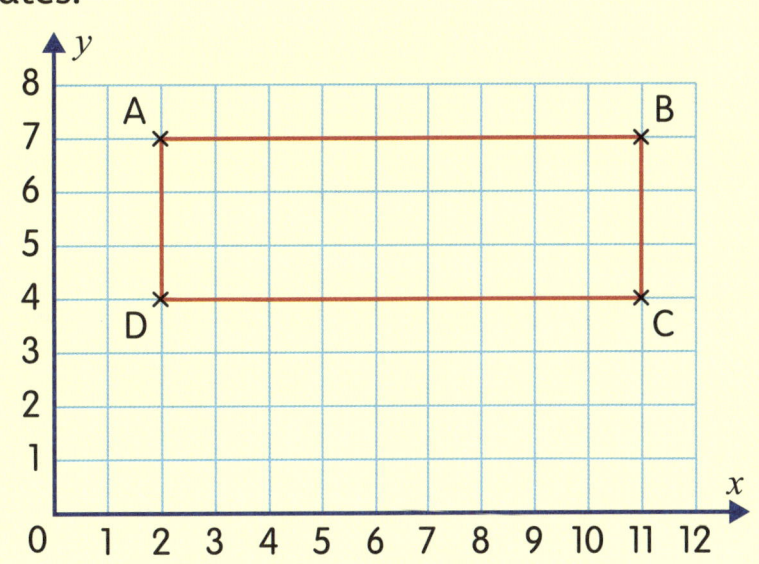

Chapter 20 Coordinates

Exercise 2

1) a) In your jotter, write the coordinates of shape ABCDE.
   A = (12, 12), B = (__, __), C = (__, __), D = (__, __), E = (__, __)

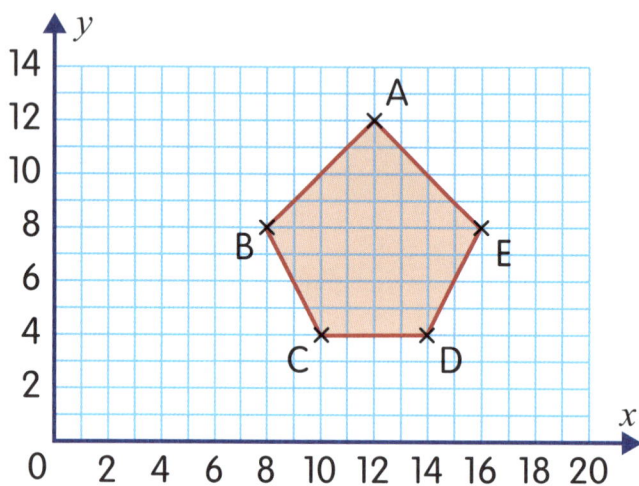

   b) What is the name of shape ABCDE?

2) a) In your jotter, write the coordinates of shape ABCDEF.

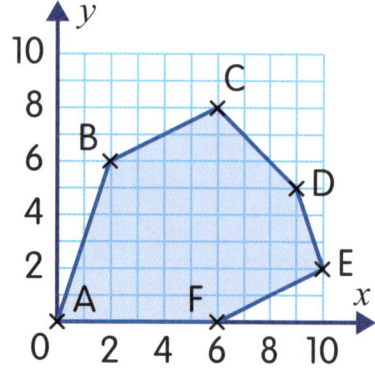

   b) What is the name of shape ABCDEF?

The coordinate axes

3) DEFG is a rectangle.
What are the coordinates of G?

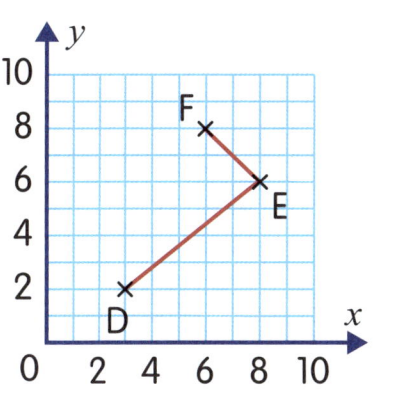

4) EFG is a triangle.

In your jotter, write a possible coordinate for E if EFG is a:
a) right-angled triangle
b) isosceles triangle
c) scalene triangle.

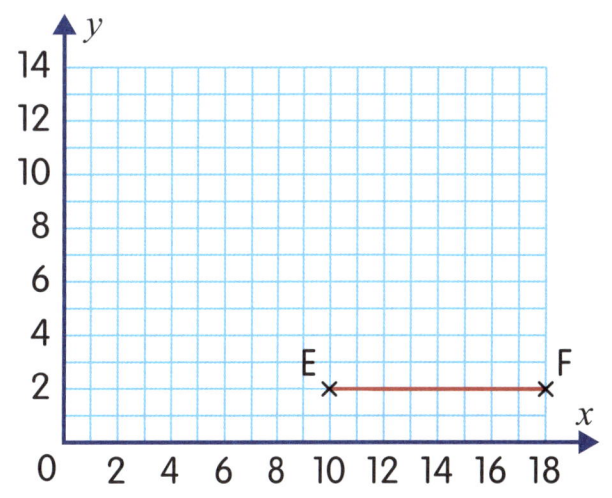

5) ABCD is a square.
AB is one side of the square.

In your jotter, write **two different** pairs of coordinates for the positions of C and D.

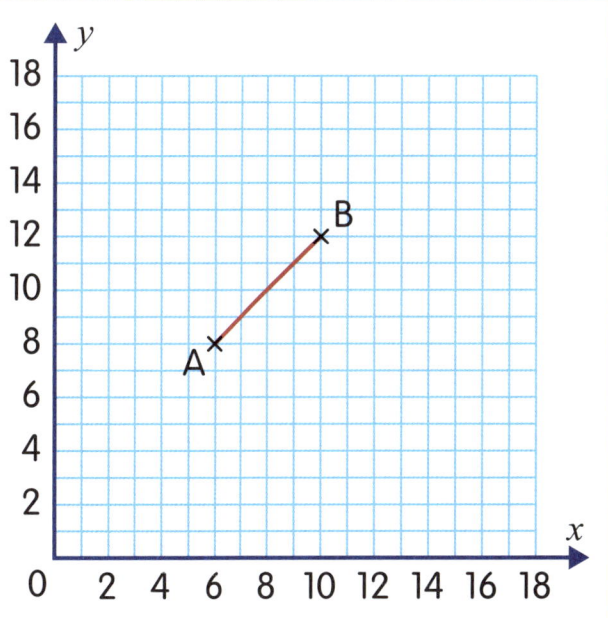

Chapter 20 Coordinates

 6) DEFG is a rectangle.

The rectangle has perimeter 20 units.

DE is one side of the rectangle.

What are the coordinates of **F** and **G**?

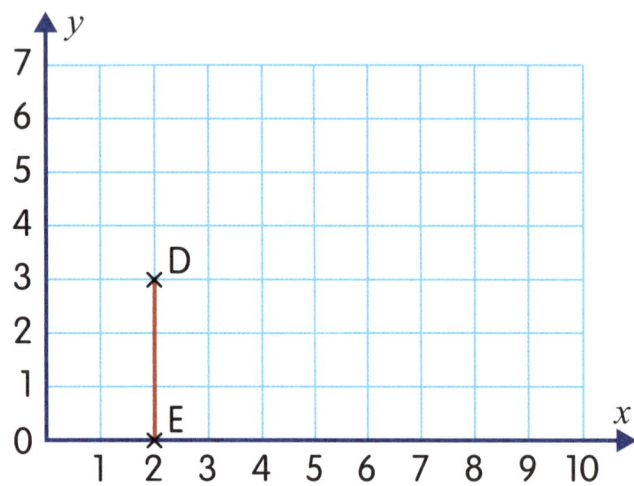

Revisit, review, revise

1) a) Draw a coordinate grid with 20 as the highest x- and y-coordinates.

b) Plot the points in order and join them as you go.

   i) (2, 3) (6, 10) (2, 12) (2, 3)

   ii) (12, 12) (17, 17) (14, 20) (9, 15) (12, 12)

   iii) (6, 4) (8, 8) (10, 7) (13, 9) (10, 1) (6, 4)

c) Write the name of each shape.

2) a) In your jotter, write the coordinates of shape ABCDEF.
   b) What is the name of shape ABCDEF?

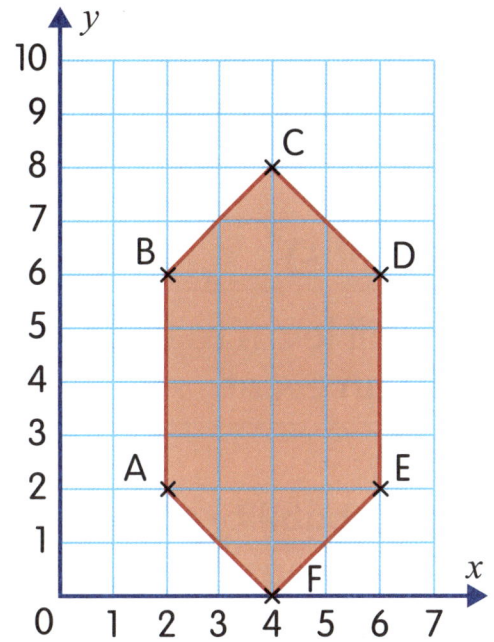

3) ABCD is a rectangle.
   What are the coordinates of D?

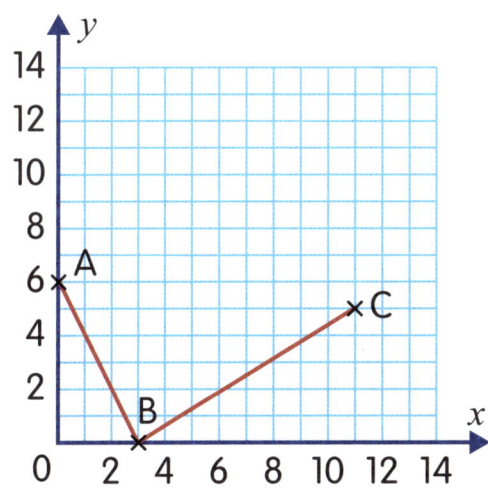

21 Probability
Understanding and predicting probability

Probability

 I will learn to understand probability and predict the probability of an event.

The **probability** of an event happening is the mathematical chance of it happening.

We can use words to describe the probability of an event.
- **Certain** – it will always happen.
- **Likely** – it will probably happen but may not.
- **An even chance** – there is the same chance of it happening as it not happening.
- **Unlikely** – it will probably not happen, but it may.
- **Impossible** – it will not happen.

Example

What is the probability that everyone in the class will like the same flavour of ice lolly?

It is **unlikely**.

It may happen but probably not!

Understanding and predicting probability

Exercise 1

Use the words from the box to answer questions 1–4.

| impossible | unlikely | even chance | likely | certain |

1) Will places cards numbered 1–8 face down on the table.
   He picks a card at random.
   What is the probability that he picks:
   a) 3
   b) a number from 1 to 6
   c) 15?

2) Ali rolls a six-sided dice.
   What is the probability he rolls:
   a) a 6   b) an even number   c) an odd number
   d) a 9   e) a number from 1 to 6?

3) A bag contains 1 strawberry, 3 orange and 4 lemon sweets.
   Azra picks one without looking.
   What is the probability that she picks:
   a) a strawberry sweet
   b) a lemon sweet
   c) a raspberry sweet?

4) Charley's goal-shooting average in basketball is 80 out of 100.
   What is the probability that Charley scores the next time she shoots?

Chapter 21 Probability

 5) Lucy is playing a game.
She rolls a six-sided dice.
She wins if she rolls a 5 or a 6.
Will she win or lose more often?

 6) Imran and Fran are deciding who should go first in a game.
Fran says: 'Let's roll a dice; if the score is lower than 3, you can go first.'
Is this fair?
Explain your answer.

Now try this!

Look at these two dice. If you add together the number of dots on them, the total is 2.

When you roll two ordinary six-faced dice and add together the two numbers, what results can you get?

Do you have more chance of getting one total than any other?

If so, what is that answer? And why?

What if you use a dice with 10 numbers on, like these?

What answers can you get now?

Do you have more chance of getting one total than any other?

If so, what is that answer? And why?

Understanding and predicting probability

Revisit, review, revise

1) Ari rolls a 10-sided dice.

   In your jotter, write an outcome whose probability is:

   a) impossible
   b) certain
   c) an even chance
   d) likely
   e) unlikely.

2) In a game, cards with the letters forming the word MATHEMATICS are placed face down on a table.

   A player picks a card at random.
   They win if they pick a vowel.
   Are players more likely to win or lose?

22 End-of-year revision

1) What percentage of the large square is:
   a) coloured
   b) **not** coloured?

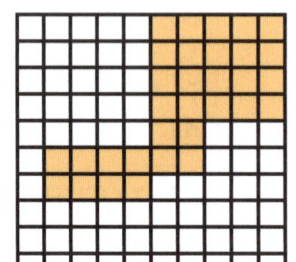

2) In your jotter, write the first three (non-zero) multiples of 9.

3) In your jotter, write all the factors of 20.

4) Esme buys a painting for £245.
   She sells it for £425.
   a) Does Esme make a profit or loss?
   b) How much profit or loss does she make?

5) Rohan is saving for a bicycle.
   The bicycle costs £675.
   He saves £105 every month for 6 months.
   How much more money does he need to buy the bicycle?

6) In the number **652 410** what do these digits represent?
   a) 2   b) 4   c) 6   d) 5

7) In the decimal **482.136** which digit represents:
   a) hundredths   b) thousandths   c) hundreds?

End-of-year revision

8) In your jotter, write **sixteen thousand, five hundred and one** in digits.

9) In your jotter, write **301 299** in words.

10) In your jotter, write these numbers in order of size. Start with the **smallest**.
   a) 105 471   104 705   105 714   14 750
   b) 76.315   76.513   76.33   76.5

11) Round:
   a) 56 312 to the nearest 10 000
   b) 899 001 to the nearest 100 000
   c) 314 876 to the nearest 10 000
   d) 4.63 to 1 decimal place
   e) 30.09 to 1 decimal place
   f) 8.657 to 2 decimal places
   g) 0.075 to 2 decimal places
   h) 4071.934 to 1 decimal place.

12) Work out:
   a) 9.28 × 10
   b) 24.3 × 100
   c) 16.2 ÷ 10
   d) 587.1 ÷ 10
   e) 8572 ÷ 1000
   f) 29.5 ÷ 100

13) Mentally work out:
   a) 4.6 + 1.3
   b) 12.7 − 0.5

14) Line up the digits with the same place value to work out:
   a) 16.8 + 2.7
   b) 478.3 − 291.4

Chapter 22 End-of-year revision

15) 28 517 people buy a magazine in March.

41 390 people buy the same magazine in April.

Estimate how many people buy the magazine in March and April.

16) Work out:

a) $\frac{1}{9}$ of 45

b) $\frac{5}{9}$ of 45

c) $\frac{3}{5}$ of 30

17) A brass band has 28 musicians.

a) $\frac{1}{7}$ play the trombone.

How many play the trombone?

b) $\frac{4}{7}$ play the saxophone

How many play the saxophone?

18) Which pairs of fractions are equivalent?

| $\frac{4}{6}$ | $\frac{8}{24}$ | $\frac{1}{3}$ | $\frac{12}{18}$ |

19) In your jotter, write each percentage as a fraction in its simplest form.

a) 17 %

b) 50 %

c) 24 %

20) In your jotter, write each fraction with a denominator of 100 and then as a percentage.

a) $\frac{3}{10}$

b) $\frac{1}{5}$

c) $\frac{2}{25}$

d) $\frac{7}{50}$

21) In your jotter, write each division using ⟌

Then work out the answer, including any remainder.

a) 87 ÷ 4

b) 117 ÷ 9

c) 79 ÷ 6

d) 206 ÷ 8

End-of-year revision

22) Sally is buying cupcakes for a party.
They are sold in packs of 4.
She needs 62 cupcakes.
How many packs should she buy?

23) How many days between 23rd June and 4th July, including both dates?

24) How many:
- a) minutes in 4 hours
- b) seconds in $3\frac{1}{2}$ minutes
- c) hours in 3 days
- d) months in a year
- e) days in a leap year?

25) Here is a timetable for the Edinburgh to Aberdeen bus:

Edinburgh	07:50	10:20	13:20	16:50	19:35
Kinross	08:42	11:12	____	17:42	20:27
Perth	09:08	____	14:45	18:08	20:53
Dundee	09:35	12:06	15:13	18:35	21:20
Forfar	10:04	12:35	____	____	21:49
Aberdeen	11:05	13:36	16:38	20:00	22:50

- a) What time does the 16:50 bus from Edinburgh arrive in Dundee?
- b) How long does the 11:12 bus from Kinross take to get to Aberdeen?
- c) I arrive in Perth at 5:40 p.m.
How long do I have to wait for a bus to Forfar?

26) Tas runs the first lap of a cross-country race in 3 minutes 48 seconds. She runs the second lap in 4 minutes 17 seconds.
   a) What is Tas's total time for the two laps?
   b) How many seconds quicker is Tas on the first lap than the second lap?

27) Describe each sequence and write the next two terms in the sequences in your jotter.
   a) 5, 10, 15, 20 …   first term = ____   term-to-term rule = ____
   b) 91, 87, 83, 79 …   first term = ____   term-to-term rule = ____

28) In your jotter, write the next two terms in each sequence.
   a) 1, 4, 9, 16 …   b) 1, 3, 6, 10 …   c) 1, 1, 2, 3, 5, 8 …

29) These rugs have patterns of black and white squares.

1 rug   2 rugs   3 rugs
9 black squares   18 black squares   27 black squares

   a) Copy and complete the table.

No. of rugs	1	2	3	4	5	6
No. of black squares	9	18	__	__	__	__

   b) Copy and complete the formula:
      number of black squares = ____ × number of rugs
   c) Use your formula to work out how many black squares there are on 12 rugs.
   d) How many rugs have 180 black squares **altogether**?

30) a) Measure these angles and write their sizes in your jotter:

i)

ii)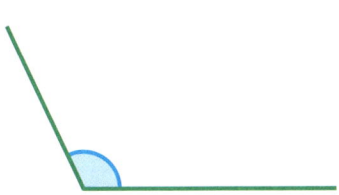

b) Is each angle in part a) acute, reflex, right or obtuse?

31) Draw and label these angles:
 a) <ABC = 53°
 b) <STU = 118°

32) Name and describe each triangle.
Choose **all** the correct names from the box.

| scalene | isosceles | equilateral | right-angled |

a) Triangle PQR (right angle at P)

b) Triangle KLM (with tick marks on KL and KM, equal angles at L and M)

c) Triangle STU (with tick marks on all three sides, equal angles at S, T and U)

33) a) On squared paper draw a rectangle ABCD with:
 • AB = 10 cm
 • AB = CD
 • perimeter = 34 cm.

b) Calculate the area of the rectangle.

34) Are these statements true or false?
 a) 14 − 7 ≠ 28 ÷ 4
 b) 32 ÷ 8 > 3 − 0
 c) 12 × 5 = 31 + 29
 d) $\frac{1}{2}$ of 60 < 7 × 4

Chapter 22 End-of-year revision

35) What does the symbol ◇ stand for in each calculation?
Choose from +, −, × or ÷
 a) 8 ◇ 6 = 14
 b) 36 = 4 ◇ 9
 c) 15 ◇ 8 = 7
 d) 6 = 30 ◇ 5

36) Work out what the ⌂ stands for in each equation.
 a) ⌂ + 7 = 15
 b) 9 − ⌂ = 4
 c) 48 = ⌂ × 6
 d) ⌂ = 30 ÷ 3

37) There are 14 adults in a group of 32 people.
In your jotter, write an equation and solve it to work out how many children are in the group.

38) a) Find the output for each function machine.
 i) 17 → × 10 → OUTPUT
 ii) 18 → − 0 → OUTPUT

b) Find the input for each function machine.
 i) INPUT → + 12 → 93
 ii) INPUT → ÷ 7 → 12

39) Copy and complete the shapes so the red dashed line is a line of symmetry each time.

a)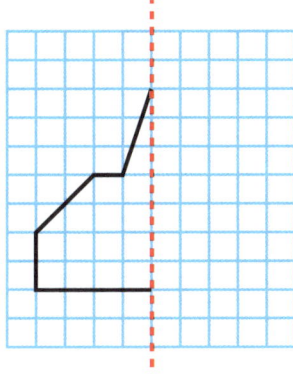

b)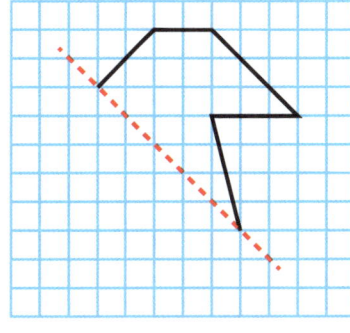

40) a) Calculate the volume of this cuboid.

b) On a piece of isometric paper, draw the cuboid.

c) On squared paper, draw a net of the cuboid.

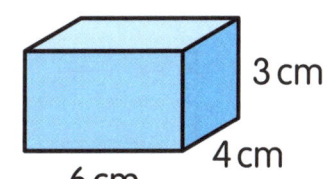

41) The comparative line graph shows the average monthly maximum and average monthly minimum temperatures (in °C) in Majorca.

a) What is the average maximum temperature in:
  i) January
  ii) July?

b) What is the average minimum temperature in:
  i) April
  ii) December?

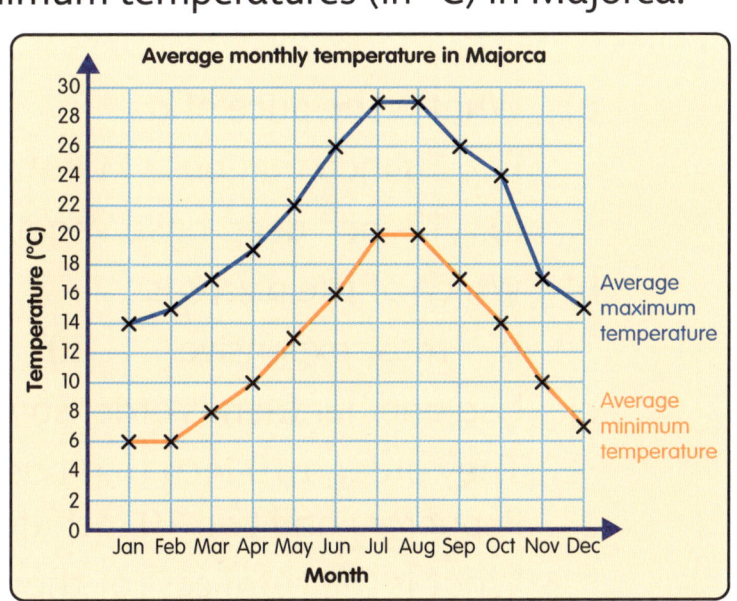

Chapter 22 End-of-year revision

c) What is the difference between the average maximum and average minimum temperatures in:

i) June   ii) November?

d) Between which two consecutive months did the average maximum temperature decrease most?

e) Between which two consecutive months did the average minimum temperature increase most?

42) The prices of pizzas from Ali's Pizzeria are shown in this table. Ali's Pizzeria offers free delivery.

	Thin base		Deep base	
	10 inch	12 inch	10 inch	12 inch
Margherita	£7.00	£9.00	£7.50	£9.60
Mushroom	£8.80	£11.00	£9.10	£11.20
Pepperoni	£9.20	£12.00	£9.60	£12.50
Vegetarian	£7.90	£10.00	£8.40	£10.60

a) What is the price of a:

i) 10-inch thin-base Margherita pizza

ii) 12-inch deep-base Vegetarian pizza?

b) Which of the pizzas is:

i) most expensive   ii) least expensive?

c) Lisa and her friends order some pizzas to be delivered.

They order a 10-inch thin-base Mushroom, a 12-inch thin-base Pepperoni and two 10-inch deep-base Vegetarian pizzas.

What is the **total** cost of the order?

43) The bar graph shows Ahmed's marks (out of 100) in three maths tests.

a) Explain why the graph is misleading.

b) Draw a bar graph that gives a better comparison of marks in December, March and June.

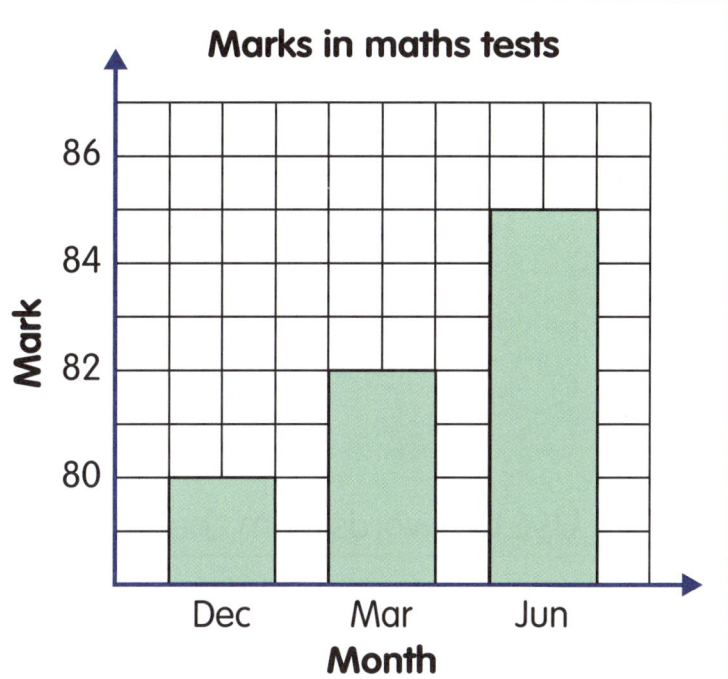

44) Draw a 10 × 10 grid as shown.

a) In your jotter, write the coordinates of:
   i) A   ii) B.

b) Plot the point C (8, 2).

c) ABCD is a rectangle. Plot the point D.

d) What are the coordinates of D?

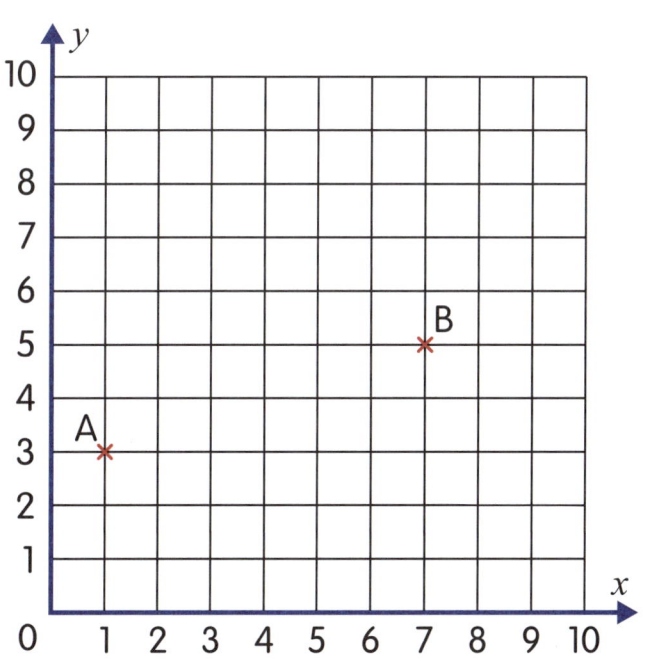

Chapter 22 End-of-year revision

45) Sula chooses one day at random from the days in November 2024.

NOVEMBER 2024						
M	T	W	T	F	S	S
				1	2	3
4	5	6	7	8	9	10
11	12	13	14	15	16	17
18	19	20	21	22	23	24
25	26	27	28	29	30	

Use the words from the box to answer the questions.

impossible unlikely even chance likely certain

What is the probability she chooses:

a) 5th November

b) a Monday, Tuesday, Wednesday or Thursday after 3rd November

c) 31st November

d) a day before 16th November

e) a day after 31st October and before 1st December?